HOUGHTON MIFFLIN HARCOURT

Math

Intensive Intervention

Teacher Guide

Grade 2

HOUGHTON
MIFFLIN
HARCOURT
School Publishers

www.hmhschool.com

Contents

Using Intensive Intervention.. v

Skills

1   Use Pictures to Add..IIN1

2   Use Symbols to Add... IIN3

3   Use Pictures to Subtract... IIN5

4   Use Symbols to Subtract... IIN7

5   Model Numbers to 20.. IIN9

6   Identify Numbers to 30.. IIN11

7   Compare Numbers to 30... IIN13

8   Order on a Number Line... IIN15

9   Addition Patterns... IIN17

10 Subtraction Patterns... IIN19

11 Use a Number Line to Count On.. IIN21

12 Add Doubles... IIN23

13 Doubles Plus One... IIN25

14 Add in Any Order... IIN27

15 Model Adding Tens... IIN29

16 Use a Number Line to Count Back.. IIN31

17 Related Subtraction Facts to 12... IIN33

18 Fact Families to 12... IIN35

19 Think Addition to Subtract... IIN37

20 Model Subtracting Tens... IIN39

21 Read a Tally Table... IIN41

22 Read a Picture Graph.. IIN43

23 Make a Prediction.. IIN45

24	Pennies, Nickels, and Dimes	IIN47
25	Skip-Count by Fives and Tens	IIN49
26	Count Money	IIN51
27	Pennies, Nickels, Dimes, and Quarters	IIN53
28	Use a Clock	IIN55
29	Minutes and Hours	IIN57
30	Sort by Color, Size, and Shape	IIN59
31	Identify Three-Dimensional Shapes	IIN61
32	Sides and Vertices	IIN63
33	Sort Two-Dimensional Shapes	IIN65
34	Identify and Copy Patterns	IIN67
35	Extend Patterns	IIN69
36	Compare Lengths	IIN71
37	Temperature	IIN73
38	Compare Weights	IIN75
39	Compare Capacities	IIN77
40	Equal Parts	IIN79
41	Explore Halves	IIN81
42	Explore Thirds and Fourths	IIN83
43	Explore Place Value to 50	IIN85
44	Explore Place Value to 100	IIN87
45	Compare Numbers to 50	IIN89
46	Order Numbers to 100 on a Number Line	IIN91
47	Model Addition with 1-Digit and 2-Digit Numbers	IIN93
48	Mental Math: Add Tens	IIN95
49	Model Subtraction with 2-Digit and 1-Digit Numbers	IIN97
50	Mental Math: Subtract Tens	IIN99
51	Skip-Count by Twos and Fives	IIN101
52	Skip-Count on a Hundred Chart	IIN103

Using Intensive Intervention

Intensive Intervention is targeted at children who are performing two or more years below grade level. By focusing on essential prerequisite skills and concepts, *Intensive Intervention* prescribes instruction to prepare children for grade-level success in your mathematics program.

How do I determine if a child needs *Intensive Intervention*?

Use the Show What You Know pages at the beginning of each chapter of the Student Edition to diagnose a child's need for intervention. Show What You Know targets the prerequisite skills necessary for success in each chapter. If a child misses a limited number of exercises, that child would benefit from *Strategic Intervention. Strategic Intervention* is prescribed in the Teacher Edition for each chapter. Children who miss at least half of the exercises from Show What You Know are candidates for *Intensive Intervention*. Use the prescription chart in the *Intensive Intervention User Guide* to determine whether or not a child requires *Intensive Intervention*.

Which *Intensive Intervention* skill lessons do I assign for each chapter?

The Chapter Correlation in the *Intensive Intervention User Guide* correlates the *Intensive Intervention* skills to each chapter of your mathematics program.

Once you have identified a child as needing *Intensive Intervention* and have determined which skills cover prerequisites for that chapter, use the Pre-Assess activities for each recommended skill to identify the specific, intensive prerequisite skills the child needs to develop. The Pre-Assess activities, which appear at the beginning of the teacher pages for each skill lesson, will help you determine which skills will help the child succeed in each chapter.

Alternatively, some children may benefit from completing each of the skill lessons in sequence.

What materials and resources do I need for *Intensive Intervention*?

Intensive Intervention includes the Teacher Guide and Skill Packs for each grade level. The teaching strategies may require the use of common classroom manipulatives or easily gathered classroom objects. Since these activities are designed for only those children who show weaknesses in their skill development, the quantity of materials will be small. For many activities, you may substitute materials, such as paper squares for tiles, coins for two-color counters, and so on.

How are the skill lessons structured?

Each skill lesson includes two student pages and two pages of teacher support in the Teacher Guide. Each lesson begins with Learn the Math, a guided page that provides a model or an explanation of the skill. The second part of the lesson is Do the Math, a selection of exercises that provide practice and may be completed independently, with a partner, or with teacher direction. Children who have difficulty with the Do the Math exercises may benefit from the Alternative Teaching Strategy activity provided in the Teacher Guide.

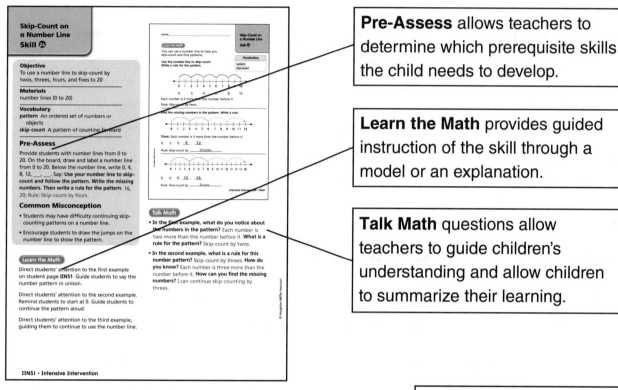

Pre-Assess allows teachers to determine which prerequisite skills the child needs to develop.

Learn the Math provides guided instruction of the skill through a model or an explanation.

Talk Math questions allow teachers to guide children's understanding and allow children to summarize their learning.

The **Alternative Teaching Strategy** provides another way for the child to acquire the skill by using a hands-on format. The activity is useful for different types of learners.

Do the Math provides practice, allowing children to demonstrate mastery of prerequisite skills.

The **Check** question allows children to demonstrate their understanding.

How can I assess children's understanding?

Use the Talk Math questions in the Teacher Guide to encourage children to verbalize their thinking and understanding as they complete the skill lesson. The Check at the end of the Do the Math section also provides children an opportunity to demonstrate their understanding and summarize their learning. These verbal and written responses allow teachers to assess children's knowledge of these prerequisite skills and determine if the children require further instruction with the Alternative Teaching Strategy.

What support is provided for English Language Learners?

The first student page of every lesson provides examples and visuals carefully selected to illustrate the mathematics and promote discussion. Words on the student pages are deliberately kept to a minimum so that all children can be successful, regardless of English language acquisition. Language support for math vocabulary is provided in the Teacher Guide for most lessons. Additional support for vocabulary development can be found online at the animated Multilingual eGlossary at **www.harcourtschool.com/hspmath**, which is also accessible through The Harcourt Learning Site.

How can I organize my classroom and schedule time for intervention?

You may want to set up a Math Skill Center with a record folder for each child. Assign appropriate skills in each child's folder, then allow children to work through the intervention materials, record the date of completion, and place the completed work in their folders for your review.

Children might visit the Math Skill Center for a specified time during the day, two or three times a week, or during free time. You may wish to assign children a partner, assign a small group to work together, or work with individuals one-on-one.

Use Pictures to Add
Skill ❶

Objective
To use pictures to add

Materials
counters

Pre-Assess
Provide children with 5 counters. Have them arrange the counters in a group of 3 and a group of 2. Ask children to join the counters and then count them to find how many counters there are in all. Repeat the activity with 2 and 1, and 2 and 2. Ask: **How many counters are there in each group? How many counters are there in all?**

Common Misconception
• Children may count one of the objects more than once while trying to find the total number of objects.

• To correct this, have children move each object to the side as it is counted. Explain that when they do this, they will know which ones they have already counted.

Learn the Math

Have children use counters to model the first problem on student page **IIN1**. Point out that they will use counters to represent the apples in the pictures.

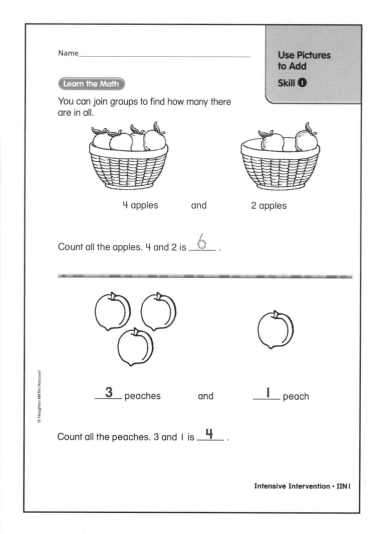

Name_____

Use Pictures to Add
Skill ❶

Learn the Math

You can join groups to find how many there are in all.

4 apples   and   2 apples

Count all the apples. 4 and 2 is __6__ .

__3__ peaches   and   __1__ peach

Count all the peaches. 3 and 1 is __4__ .

Intensive Intervention · IIN1

Talk Math

• **How many apples are there in the first basket?** 4 apples Have children set out 4 counters.

• **How many apples are there in the second basket?** 2 apples Have children set out 2 counters.

• Have children join the two groups of counters. Ask: **How many apples are there in all?** 6 apples

• **What is 4 and 2?** 6

Have children trace the number 6 with their pencils. Repeat the activity in a similar manner for the second problem.

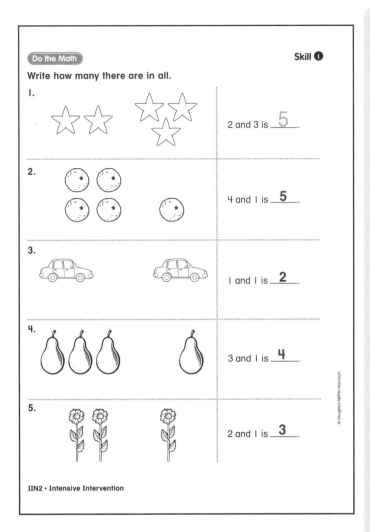

Skill ❶

Write how many there are in all.

1.
2 and 3 is ___5___.

2.
4 and 1 is ___5___

3.
1 and 1 is ___2___.

4.
3 and 1 is ___4___

5.
2 and 1 is ___3___.

© Houghton Mifflin Harcourt

Alternative Teaching Strategy

How Many Children?

Objective To act out addition with sums to 6

- Explain to children that they are going to act out addition stories using themselves as objects. Ask two volunteers to stand in front of the class.

- Have them pretend to swim. Say: **Two children are swimming.** Ask two more volunteers to join them and pretend to swim. Say: **Two more children join them. How many children are swimming in all?** Count with the class the volunteers who are acting out the swimming story. 4 children are swimming

- Repeat the activity telling different addition stories until all children have had an opportunity to volunteer.

- Challenge children to make up their own addition stories for others to act out and solve.

Do the Math

Help children see that there are two groups of objects in each problem on student page **IIN2**. They need to find how many objects there are in all. Remind them to count all of the objects to find the sum.

Talk Math

- **What two numbers do you add in Problem I?**
2 and 3

- **How can you use the picture to add 2 and 3?**
Count all of the stars in the picture.

- **What is 2 and 3?** 5

Check

Say: **Draw a picture you can use to add I and 2.** Check that children draw the same kind of object in each group. For example, they should draw I pencil in one group and 2 pencils in the other group. Ask: **What is I and 2?** 3

© Houghton Mifflin Harcourt

Intensive Intervention · IIN2

Objective
To use + and = to write addition sentences

Materials
counters

Vocabulary
plus (+)
is equal to (=)
sum
addition sentence

Pre-Assess

Provide children with 8 counters. Have them arrange the counters in a group of 5 and a group of 3. Ask children to use the counters to find 5 + 3. Say: **Write an addition sentence.** 5 + 3 = 8 Repeat the activity with different addition problems.

Common Misconception

- Children may confuse the symbols + and =.

- To correct this, remind children that + is written between the two numbers being added and = is written before the sum. Have children read their addition sentences aloud to make sure they are correct.

Learn the Math

Have children use counters to model the first problem on student page **IIN3**.

Talk Math

- **How many groups of crayons are there?** 2

- **How many crayons are there in the first group?** 3 **How many are there in the second group?** 1

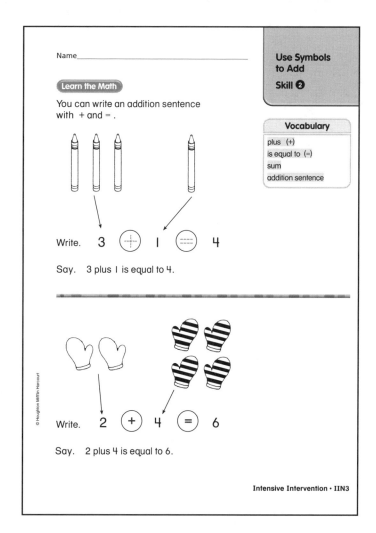

Name_____

Use Symbols to Add
Skill ❷

Learn the Math

You can write an addition sentence with + and = .

Write. 3 (+) 1 (=) 4

Say. 3 plus 1 is equal to 4.

Write. 2 (+) 4 (=) 6

Say. 2 plus 4 is equal to 6.

Vocabulary
plus (+)
is equal to (=)
sum
addition sentence

Intensive Intervention · IIN3

- **How many crayons are there in all?** 4

- **What is the addition sentence?** 3 + 1 = 4

Have students trace the + and = signs. Point out that the plus sign (+) means to add 3 and 1. The equal sign (=) tells you that 3 plus 1 more is the same as 4.

Direct children's attention to the second problem.

- **How many mittens are there in the first group?** 2 **How many are there in the second group?** 4

- **How many mittens are there in all?** 6

- **What is the addition sentence?** 2 + 4 = 6

Have students write the + and = signs.

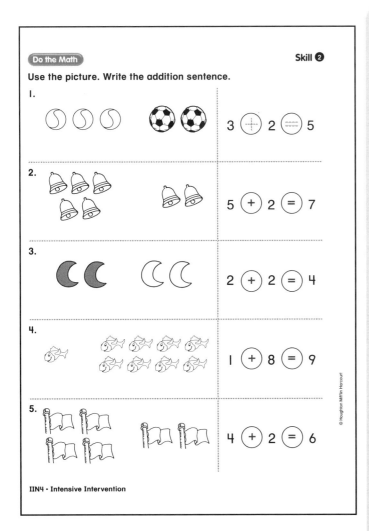

Do the Math Skill ❷

Use the picture. Write the addition sentence.

1. $3 \oplus 2 = 5$

2. $5 + 2 = 7$

3. $2 + 2 = 4$

4. $1 + 8 = 9$

5. $4 + 2 = 6$

IIN4 · Intensive Intervention

© Houghton Mifflin Harcourt

Alternative Teaching Strategy

Sentence Scramble

Objective To write addition sentences using + and =

Materials counters, paper

- Give each set of partners 10 counters.

- Write two rows of the numbers 1 through 5 on the board. Ask one partner to choose a number from the top row. Have partners write this number on a piece of paper.

- Have the other partner choose a number from the second row. Ask them to write this number on their paper.

- Say: **Add these two numbers. Use your counters to find the sum. What is the sum?** Answers may vary depending on the numbers.

- Say: **Now write an addition sentence.** Ask: **What two symbols will you use to write an addition sentence?** + and =

- Invite partners to tell their addition sentences. Ask: **What is the addition sentence?** Answers may vary depending on the numbers. Write the addition sentences on the board.

Do the Math

Have children look at student page **IIN4**. Point out that they have to complete a number sentence for each problem by writing the missing symbols. Give children counters to model the pictures.

Talk Math

- **What two symbols are missing in each addition sentence?** + and =

- **Which symbol do you write in the first circle?** +

- **Which symbol do you write in the second circle?** =

Check

Say: **Juan reads 2 books. He then reads 3 more books. What addition sentence can you write to show how many books Juan reads in all?**
$2 + 3 = 5$

Intensive Intervention · IIN4

Use Pictures to Subtract
Skill ❸

Objective
To use pictures to subtract

Materials
counters, paper

Pre-Assess

Have children draw 5 squares on a piece of paper. Then have them circle and mark an X on 3 of the squares. Ask them to use the pictures to find 5 take away 3. 5 – 3 = 2 Repeat the activity with 4 take away 1 and 3 take away 2.

Common Misconception

• Children may have difficulty visualizing the number that is taken away even though these items are circled and crossed out in the picture.

• To correct this, have them cover the number of items that are circled and crossed out with small squares of paper, so they can no longer see them. Then have them count how many are left.

Learn the Math

Have children look at the first problem on student page **IIN5**. Help children see that the circle and X over the fourth rabbit indicates that it is being taken away. You may wish to make up a subtraction story problem to go with each set of pictures.

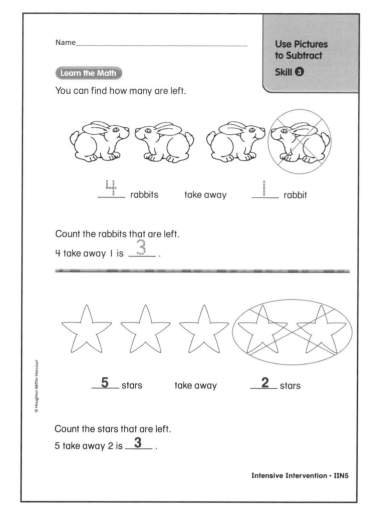

Learn the Math

You can find how many are left.

___4___ rabbits   take away   ___1___ rabbit

Count the rabbits that are left.
4 take away 1 is ___3___ .

___5___ stars   take away   ___2___ stars

Count the stars that are left.
5 take away 2 is ___3___ .

Intensive Intervention · IIN5

© Houghton Mifflin Harcourt

Talk Math

• **How many rabbits are there in all?** 4 rabbits

• **How many rabbits are taken away?** 1 rabbit

Have children trace the numbers 4 and 1.

• **How many rabbits are left?** 3 rabbits
Have children count the number of rabbits left then trace the number 3.

• **What is 4 take away 1?** 3

Ask similar questions for the second problem. As children answer each question, have them write the answer.

© Houghton Mifflin Harcourt

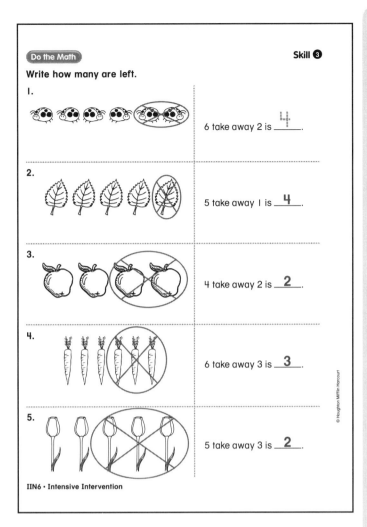

Do the Math

Write how many are left.

Skill ❸

1. 6 take away 2 is ___4___.

2. 5 take away 1 is ___4___.

3. 4 take away 2 is ___2___.

4. 6 take away 3 is ___3___.

5. 5 take away 3 is ___2___.

IIN6 · Intensive Intervention

© Houghton Mifflin Harcourt

Do the Math

Have children look at student page **IIN6**. Help them count the total number of objects in each problem and then identify the number of objects that are taken away (those that are circled and crossed out). Tell them they need to find how many are left, or not circled and crossed out.

Talk Math

- **How do you know how many to take away?**
  You take away the number of objects that are circled and crossed out.
- **How do you know how many are left?** You count the number of objects that are not circled and crossed out.

Check

Have children use counters to solve the following problem: **If you have 5 balloons and 2 blow away, how many will you have left?** You will have 3 balloons left.

Alternative Teaching Strategy

Tell a Story with Pictures

Objective To draw pictures to subtract

Materials crayons, drawing paper

- Write *4 take away 3* on the board.

- Explain to children that together they will tell a story and draw a picture to find 4 take away 3.

- Have children draw four of the same object, such as four animals, four pieces of fruit, or four toys. Ask: **How many are there to start?** 4 objects Say: **Now think of a sentence that tells about the four objects, such as "I have four oranges."** Write this sentence on the board.

- Ask: **How many should you take away?** 3 objects Have children circle and cross out three of the objects they drew. **Now think of a sentence that tells how many are taken away, such as "I give away three oranges."** Write this sentence on the board. Ask: **So, what is 4 take away 3?** 4 take away 3 is 1.

- Ask: **How many are left?** 1 object **Think of a sentence that tells how many are left. For example, "I have one orange left."** Write this sentence on the board. Ask: **So, what is 4 take away 3?** 4 take away 3 is 1.

- Repeat the activity with other subtraction problems. Encourage children to draw pictures of objects and to tell a subtraction story for each one.

Intensive Intervention · IIN6

Use Symbols to Subtract
Skill ❹

Objective
To use – and = to write subtraction sentences

Materials
counters

Vocabulary
minus (–)
is equal to (=)
subtraction sentence

Pre-Assess

Provide children with 7 counters. Have them use the counters to find 7 – 3. Ask children to record the subtraction sentence. 7 – 3 = 4 Repeat the activity with 5 – 2 and 6 – 3. Make sure children correctly identify the number to model first, the number taken away, and the number that is left.

Common Misconception

- Children may mix up the first two numbers in a subtraction sentence.

- To correct this, tell children that the first number always tells how many are in the whole group and the second number (the number after the minus symbol) tells how many are taken away.

Learn the Math

Have children use counters to model the first problem on student page **IIN7**. Have them point to each part of the subtraction sentence as they answer the questions.

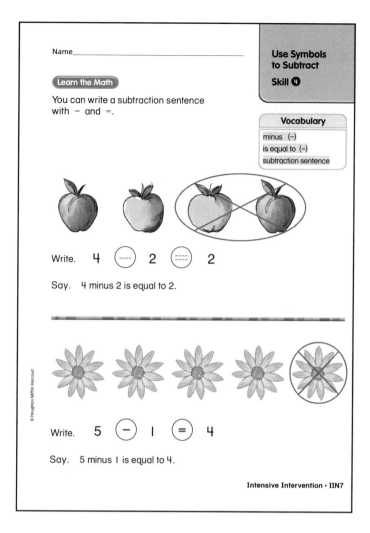

Learn the Math

You can write a subtraction sentence with – and =.

Vocabulary
minus (–)
is equal to (=)
subtraction sentence

Write.   4 (–) 2 (=) 2

Say.   4 minus 2 is equal to 2.

Write.   5 (–) 1 (=) 4

Say.   5 minus 1 is equal to 4.

Intensive Intervention · IIN7

Talk Math

- **How many counters should you begin with?** 4 counters Pause while children set out the counters.

- **How many counters should you take away?** 2 counters **How do you know?** 2 apples are circled and crossed out. Have children remove 2 counters.

- **How many counters are left?** 2 counters

- **Do you have more or fewer counters now?** fewer counters

Point out the minus and equal signs in the subtraction sentence and have children trace them with their pencils. Repeat the modeling and questions for the second problem.

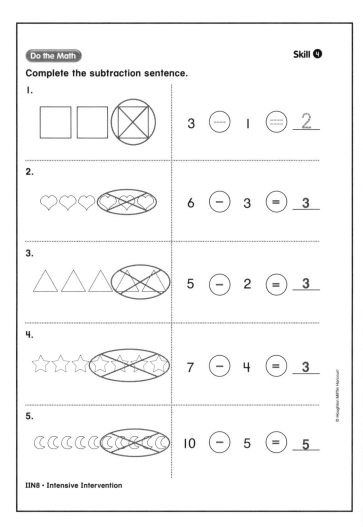

Complete the subtraction sentence.

1. 3 ⊖ 1 ⊜ 2

2. 6 ⊖ 3 ⊜ 3

3. 5 ⊖ 2 ⊜ 3

4. 7 ⊖ 4 ⊜ 3

5. 10 ⊖ 5 ⊜ 5

IIN8 • Intensive Intervention

Do the Math

Have children look at student page **IIN8**. Encourage them to use counters to model each problem.

Talk Math

• **What two symbols are missing in each subtraction sentence?** – and =

• **What do these symbols mean?** minus and is equal to

• **How do you know how many to take away?** You count the number of objects that are circled and crossed out.

• **How can you tell how many are left?** You count the number of objects that are not circled and crossed out.

Check

Ask: **What are the symbols for *minus* and *is equal to*?** – and =

Alternative Teaching Strategy

How Many Are Left Standing?

Objective To act out subtraction stories and read subtraction sentences

• Tell the class the following story:

   There are 6 children.
   2 walk away.
   How many children are left?

• Have children act out the story. Ask six volunteers to stand in a line, facing the class. Ask: **How many children are there?** 6 children

• Have the last two children in the row return to their seats. Ask: **How many children returned to their seats?** 2 children

• Write 6 – 2 = _____ on the board. Point out the minus and equal signs.

• Ask: **How many children are left?** 4 children **How can you complete the subtraction sentence?** Write 4 on the blank space.

• Encourage children to read the subtraction sentence along with you. Say: **Six minus two is equal to four.**

• Repeat the activity with other subtraction stories. You may wish to have children tell their own stories for the class to act out.

Objective
To model numbers to 20

Materials
counters, ten frames

Pre-Assess

Model the number 8 using counters and a ten frame. Ask children to tell what number is shown. Next, tell children you want to show the number 5 with a ten frame and counters. Ask children to tell how many counters you need. Repeat with other numbers including numbers between 1 and 10.

Common Misconception

• Children may forget to fill the first ten frame with 10 counters before moving on to the second ten frame when modeling numbers from 11 to 20.

• To correct this, remind children that they must completely fill the first ten frame with ten counters before they place counters in the second ten frame. Point out that if they do not fill the first ten frame before moving on to the second, they may show an incorrect number. Remind children to always fill the ten frame from left to right, top to bottom.

Learn the Math

Have children use counters and a ten frame to model the first problem on student page **IIN9**.

Learn the Math

You can use counters to show numbers.

How can you show 6 with counters? Draw counters to show 6.

Draw counters to show 9.

Draw counters to show 14.

Intensive Intervention • IIN9

Talk Math

• **What number are you asked to show?** 6

• **How many counters will you use?** 6 counters

Count each counter in the first problem aloud. Have children trace the counters. Repeat the questions for the second problem. Provide children with two ten frames to model the third problem. Check that they completely fill the first ten frame before moving on to the second ten frame.

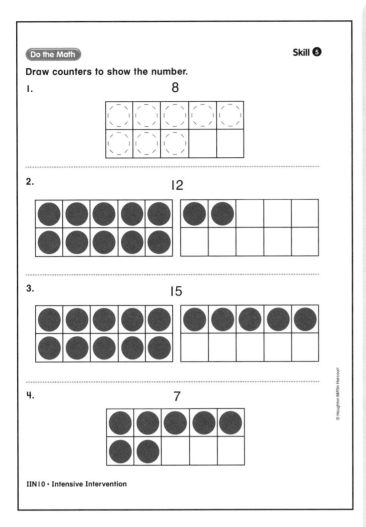

Do the Math

Have children look at student page **IIN10**. Point out that there are 10 boxes in a ten frame. Explain that if a number is greater than 10, they should draw the first 10 counters in the first ten frame and the rest of the counters in the second ten frame.

Talk Math

- **How do you know how many counters to draw in each problem?** You draw the same number of counters as the number you want to show.

- **How many ten frames do you need to show a number between 11 and 20?** 2 ten frames

Check

Ask: **How many ten frames will you need to show 7 counters?** I ten frame **Will any boxes in the ten frame be empty?** yes **How many?** 3 boxes

Alternative Teaching Strategy

Picture That!

Objective To model numbers to 20 using pictures and counters

Materials counters, ten frames, crayons, drawing paper

- Draw 4 stars on the board. Ask: **How many stars are there?** 4 stars **So, what number do they show?** 4

- Say: **I just showed the number 4 with a picture.** Ask: **How can I show the number 4 with counters?** Show 4 counters.

- Draw an empty ten frame on the board. Ask a volunteer to draw 4 counters in the ten frame.

- Tell children they will practice showing numbers with counters. Have them draw several of the same object such as bananas or flowers. Ask them to draw any number of the same object up to 20.

- Ask: **How many does your picture show?** Answers may vary. Say: **Now use your counters and ten frames to show the same number.**

- Repeat several times so children can practice showing different numbers with pictures and counters. Point out that a number can be modeled by using pictures or counters.

Identify Numbers to 30
Skill ❻

Objective
To identify numbers and count up to 30 objects

Materials
counters

Pre-Assess
Draw 10 squares on the board in two rows of five. Ask children to tell how many squares there are. Put a box around the 10 squares. Draw another group of 10 squares in a similar way. Ask children to tell how many squares there are now. Then draw 6 more squares to the right of the second group of 10. Ask children to tell how many squares there are in all. Check to see if children count groups of 10 first and then count on the 6 additional squares.

Common Misconception
- Children may continue to count all of the objects in a large set of objects, even when the objects are organized in groups of ten.
- To correct this, remind them to count by tens. Point out that counting by tens saves time and there is less chance they will make a mistake.

Learn the Math
Guide children through the first problem on student page **IINII**.

Learn the Math

You can use groups of 10 to find how many.

How many ☽ are there?

Count groups of 10 first.

How many groups of 10 ☽ are there? __2__		2 tens = 20
How many other ☽ are there? __4__		4 ones = 4
How many ☽ are there in all? __24__		2 tens 4 ones = 24

How many [ERASER] are there?

Count the group of 10 first.
Then count the rest.

There are __15__ [ERASER] in all.

1 ten =	__10__
5 ones =	__5__
1 ten 5 ones =	__15__

© Houghton Mifflin Harcourt

Intensive Intervention · IINII

Talk Math

- **How many moons are in the first group?** 10
- **How many moons are there in the second group?** 10
- **How many groups of ten are there?** 2 **How much is 2 tens?** 20
- **How many other moons are there?** 4
- **How many moons are there in all?** 20 and 4 more, or 24

Have children trace the answers to the questions in the first problem. Repeat the questions for the second problem.

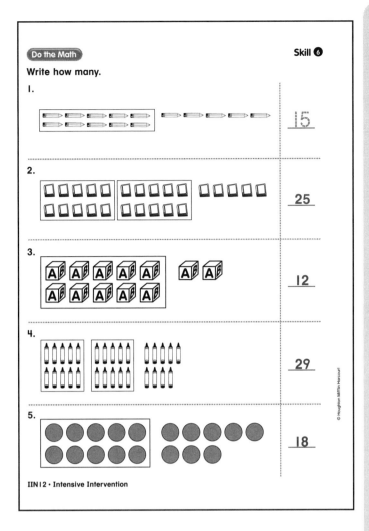

Write how many.

1.

    15

2.

    25

3.

    12

4.

    29

5.

    18

IIN12 • Intensive Intervention

© Houghton Mifflin Harcourt

Alternative Teaching Strategy

Stacks of 10

Objective To count numbers in groups to 30

- Give pairs of children 30 counters. Ask them to stack 10 counters on top of each other. Ask: **How many are in the stack?** 10

- Have children make another stack of 10 counters next to the first. Say: **Count by tens to find the total number of stacked counters.** Ask: **How many counters are stacked?** 20

- Have children place 3 counters next to their group. Ask: **Now how many counters are there in all?** 23

- Have one child in the pair create a group of counters between 11 and 30 by stacking groups of ten and placing the rest next to the stacks. Have the other child in the pair determine the total number of counters. Then have children switch roles. Have them repeat several times to practice counting by tens and ones to 30.

Do the Math

Have children look at student page **IIN12**. Point out that each boxed group of objects is a group of 10. Tell children to first count the groups of ten and then count the rest of the objects to find the total number of objects.

Talk Math

- **Why is it easier to count by tens than to count each object one by one?** It is faster.

- **How many groups of ten are in Problem 1?** 1 **In Problem 2?** 2

Check

Say: **Kyra has 10 marbles in a drawer. She has 3 more marbles on her desk.** Ask: **How many marbles does Kyra have in all?** 13

© Houghton Mifflin Harcourt

Compare Numbers to 30
Skill ❼

Objective
To compare numbers to 30

Materials
base-ten blocks

Vocabulary
tens
ones

Pre-Assess

Use base-ten blocks to model 14 and 17. Show children the model for 14. Say: **This shows 14.** Ask: **How many tens are there? 1 How many ones are there?** 4 Show children the model for 17. Say: **This shows 17.** Ask: **How many tens are there? 1 How many ones are there?** 7 **Which number is greater: 14 or 17?** 17

Common Misconception

• Children may assume that if a number has more ones than another number, it is greater than the other number.

• To correct this, remind children that they first need to compare the tens. The number with more tens is greater. If the tens are the same, then they need to compare the ones. Point out that if a number has only ones blocks, it has zero tens.

Learn the Math

Have children model the first problem on student page **IIN13** using base-ten blocks. Work through the problem together.

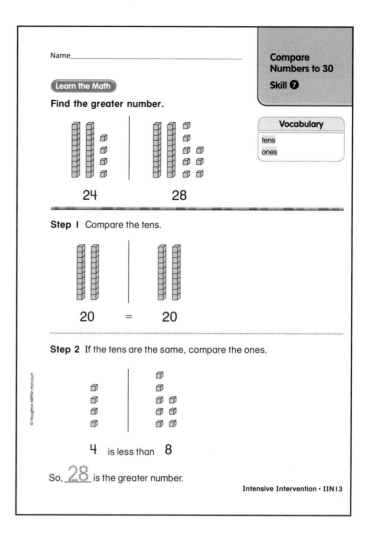

Talk Math

• **How many tens are there in 24?** 2

• **How many tens are there in 28?** 2

• **How many ones are there in 24?** 4

• **How many ones are there in 28?** 8

• **Which is greater: 24 or 28?** 28

Have children trace the greater number.

• **What if you were asked to find the lesser number? Which is the lesser number: 24 or 28?** 24

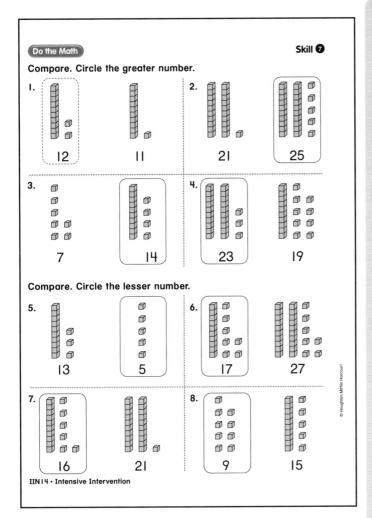

Do the Math

Skill **7**

Compare. Circle the greater number.

1. 12 11
2. 21 **25**
3. 7 **14**
4. **23** 19

Compare. Circle the lesser number.

5. 13 **5**
6. **17** 27
7. **16** 21
8. **9** 15

IIN14 • Intensive Intervention

© Houghton Mifflin Harcourt

Do the Math

Have children look at student page **IIN14**. Point out that in Problems 1–4, children are asked to circle the greater number and in Problems 5–8, children are asked to circle the lesser number. Have children model each problem with base-ten blocks as you work through the problems together.

Talk Math

- **Which do you compare first: the tens or the ones?** the tens

- **What should you do if the tens are the same?** Compare the ones.

Check

Ask: **Why do you need to compare the tens before the ones?** The tens have greater value than the ones. The number with more tens is greater.

Alternative Teaching Strategy

Which is Greater?

Objective To compare numbers to 30

Materials pencils, paper, chalkboard, chalk, base-ten blocks

- Draw a place-value chart like the one below on the board.

Tens	Ones

- Write the numbers 16 and 12 on the board. Explain to children that they will use the place-value chart and base-ten blocks to compare the numbers.

- Have children model each of the numbers using base-ten blocks. Ask: **How many tens are there in 16?** 1 Write 1 in the first row of the tens column. **How many ones are there in 16?** 6 Write 6 in the first row of the ones column.

- Ask: **How many tens are there in 12?** 1 Write 1 in the second row of the tens column. **How many ones are there in 12?** 2 Write 2 in the second row of the ones column.

- Compare the tens with children. Ask: **Are the tens the same?** yes Say: **Compare the ones.** Ask: **Which is greater: 6 or 2?** 6 **Which is the greater number: 16 or 12?** 16

- Repeat the activity with other pairs of numbers up to 30.

© Houghton Mifflin Harcourt

Order on a Number Line
Skill ❽

Objective
To use a number line to identify the number before, between, or after a given number

Materials
blank number lines

Vocabulary
before
between
after
number line

Pre-Assess

Draw and label a number line from II to 20 on the board. Place a box around the number I2 and the number I4. Ask: **What number comes between I2 and I4?** I3 **What number comes just before I2?** II **What number comes just after I4?** I5

Common Misconception

• Children may confuse before and after.

• To correct this, remind them that a number that comes before another number must be less than the number or to the left of the number. A number that comes after another number must be greater than the number or to the right of a number. Write 4, 5, 6 on the board. Review with children which number comes just before 5 and which number comes just after 5.

Learn the Math

Have children look at the number line on student page IINI5.

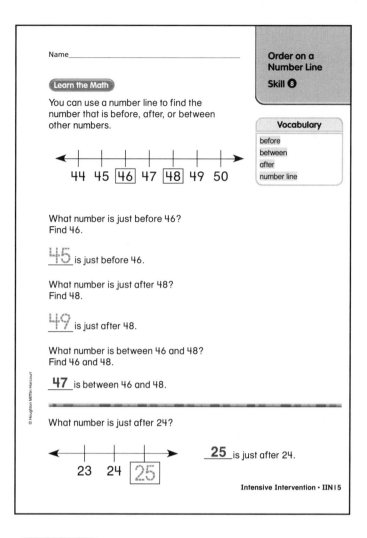

Talk Math

• **Is the number just before 46 less than or greater than 46?** less than

• **What number is just before 46?** 45

• **Is the number just after 48 less than or greater than 48?** greater than

• **What number is just after 48?** 49

• **What number is between 46 and 48?** 47

Have children circle the number on the number line that answers the first question on the student page. Then have them trace the number. Repeat the process for the remaining two questions. Guide children to complete the number line problem at the bottom of the page. Have them trace the number in the box and then write the number to complete the sentence.

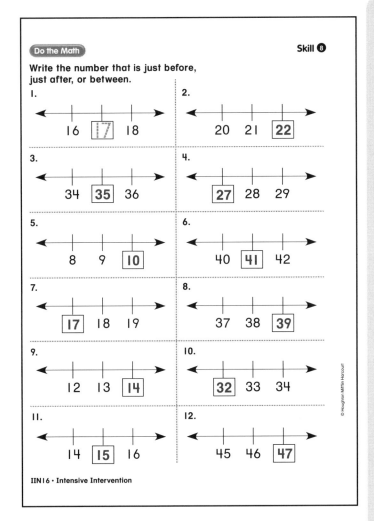

Do the Math

Have children look at student page IIN16. Point out that they need to find the number that goes in the empty box. The number is either just before, just after, or between the other numbers on the number line.

Talk Math

- **How do you know if the number you need to find is just before, just after, or between?** I look to see where the empty box is.

- **Is a number that is just before a number to the right or left of the number?** to the left

- **Is a number that is just after a number to the right or left of the number?** to the right

Check

Ask: **What number is between 27 and 29?** 28

Alternative Teaching Strategy

Which Number is Hiding?

Objective To use a number line to identify the number before, between, or after a given number

Materials number lines from 0 to 30, counters or small paper squares

- Give pairs of children a number line from 0 to 30. Give each pair of children a counter or small paper square that is large enough to cover one and only one number on the number line.

- Have children place their counter on the number 25. Ask: **What number is just before 26?** 25 **What number is just after 24?** 25 **What number is between 24 and 26?** 25

- Repeat the activity and series of questions with other numbers on the number line. Then have one partner secretly cover a number on the number line and ask his or her partner questions similar to the ones above. Have partners switch rolls and repeat the activity as time allows.

Addition Patterns
Skill ❾

Objective
To show addition patterns of one and two more in addition sentences

Materials
counters

Vocabulary
pattern

Pre-Assess

Draw one star on the board. Ask: **How many stars are there?** I star Draw one star next to the first. Under the stars, write I + I = ____. **What number completes the addition sentence?** 2 Below the first addition sentence, draw 2 stars. **How many stars are there?** 2 stars Then draw one star next to the first two. Under the stars, write 2 + I = ____. **What number completes the addition sentence?** 3 **What patterns do you see?** Possible answers: You are adding one more each time. The sum increases by one each time.

Common Misconception

• Children may not see the addition pattern when they add one to a number.

• To correct this, write these addition facts on the board, one beneath the other. Point out the patterns: one is added each time; the sum is one more each time.

$$I + I = 2$$
$$2 + I = 3$$
$$3 + I = 4$$
$$4 + I = 5$$

Work through the first set of addition sentences on student page **IINI7** with children. Have children use counters to model the first addition sentence. Encourage them to show each subsequent addition sentence by adding one more counter each time.

IINI7 • Intensive Intervention

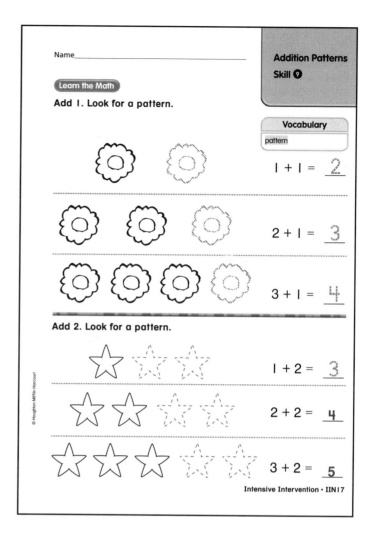

Name_____

Addition Patterns
Skill ❾

Learn the Math

Add I. Look for a pattern.

Vocabulary
pattern

$$I + I = 2$$

$$2 + I = 3$$

$$3 + I = 4$$

Add 2. Look for a pattern.

$$I + 2 = 3$$

$$2 + 2 = 4$$

$$3 + 2 = 5$$

Intensive Intervention • IINI7

© Houghton Mifflin Harcourt

Talk Math

• **In the first set of addition sentences, how much are you adding to each number?** You are adding I to each number.

• **What patterns do you see?** Possible answers: The first number in each number sentence is one more each time. The sum is one more each time.

Have children trace the sums in the first set of number sentences. Repeat the questions and activity for the second set of addition sentences in which children are asked to add 2 to a number.

© Houghton Mifflin Harcourt

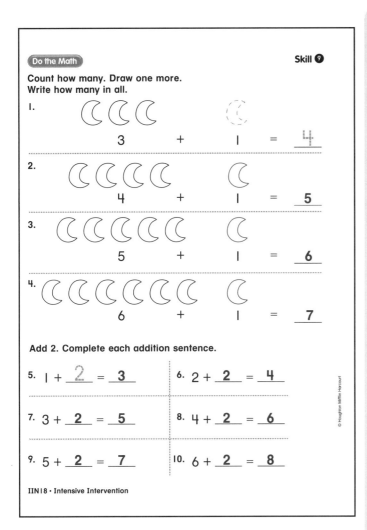

Do the Math

Skill ❾

Count how many. Draw one more.
Write how many in all.

1.
   $3 + 1 = 4$

2.
   $4 + 1 = 5$

3.
   $5 + 1 = 6$

4.
   $6 + 1 = 7$

Add 2. Complete each addition sentence.

5. $1 + 2 = 3$

6. $2 + 2 = 4$

7. $3 + 2 = 5$

8. $4 + 2 = 6$

9. $5 + 2 = 7$

10. $6 + 2 = 8$

IIN18 · Intensive Intervention

Do the Math

Have children look at student page **IIN18**. Tell them to complete each addition sentence. Encourage them to use counters to model each problem.

Talk Math

• **What pattern do you find when adding 1 more?** Possible answer: The sum is always one more than the first number.

• **What pattern do you find when adding 2 more?** Possible answer: The sum is always two more than the first number.

Check

Write the following number sentences on the board:

$8 + 1 = 9$ $8 + 2 = 10$ $8 + 3 = 11$

Ask: **How can you use a pattern to find 8 + 4 ?** Possible answer: The pattern shows that the sum will be one more each time.

Alternative Teaching Strategy

Patterns in a Chart

Objective To understand addition patterns of one more and two more

• Draw a blank two-column table with ten rows on the board.

• Explain to children that we will fill in the table with addition sentences that follows the pattern *add 1*.

• In the first column of the first row, write: 0 + 1. Point to the second column. Say: **We will write the sums in this column.** Ask: **What is the sum of 0 + 1?** 1 Write 1 in the second column of the first row.

• In the second row, write 1 + 1. Ask: **What is the sum of 1 + 1?** 2 Then write 2 + 1. **What is the sum of 2 + 1?** 3 Continue until the table is completed up to 6 + 1.

0 + 1	1
1 + 1	2
2 + 1	3
3 + 1	4
4 + 1	5
5 + 1	6
6 + 1	7

• Point to the numbers in the second column: 1, 2, 3, 4, 5, 6, and 7. Ask: **What do you notice about these numbers?** Possible answer: Each number is one more than the last number. **What do you think the next number in the column will be?** 8 Point out that the pattern of adding one more is that the sum is always one more than the first number.

• Help children continue their tables to 9 + 1. Then repeat the activity by adding 2 to the numbers 0 through 9. Discuss with children the pattern of adding 2.

Intensive Intervention • IIN18

Objective
To use a pattern to subtract

Materials
counters

Pre-Assess

Draw ten squares on the board. Ask: **How many squares are there?** 10 squares Circle and mark an X over the last square in the row. **Subtract I square.** Write 10 – 1 = ___ below the squares. **What is the difference?** 9 Erase the square with the circle and X on it. Circle and mark an X over the last square in the row. Write 9 – 1 = ___ on the board under 10 – 1 = 9. **What is the difference?** 8 Continue until you have illustrated through 6 – 1. **Do you see a pattern?** Possible answers: Each difference is one less than the difference before. The difference is one less than the number you subtract from. **If I continue the pattern, what will the next number sentence be?** 5 – 1 = 4

Common Misconception

- Children may have difficulty recognizing a pattern.

- To correct this, show them the different patterns they can find when subtracting. Point out that the difference is one less than the difference before it and that the difference is always one less than the number they started with.

Learn the Math

Guide children through each subtraction sentence on student page **IIN19**. Provide counters for children to model each problem.

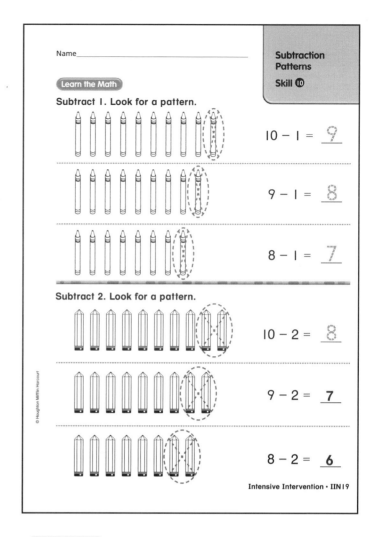

Name_____

Subtraction Patterns
Skill ⑩

Learn the Math

Subtract I. Look for a pattern.

10 – 1 = __9__

9 – 1 = __8__

8 – 1 = __7__

Subtract 2. Look for a pattern.

10 – 2 = __8__

9 – 2 = __7__

8 – 2 = __6__

Intensive Intervention · IIN19

Talk Math

- **How many counters do you begin with?** 10 counters Pause while children set out the counters.

- **How many counters do you take away?** I counter Have children remove I counter.

- **How many counters are left?** 9 counters

- **Take away I counter from 9 counters. How many counters are left?** 8 counters

- **Take away I counter from 8 counters. How many counters are left?** 7 counters

- **What do you notice about the difference each time?** Possible answer: It is one less than the difference before it.

Repeat the questions and modeling for the last set of problems.

© Houghton Mifflin Harcourt

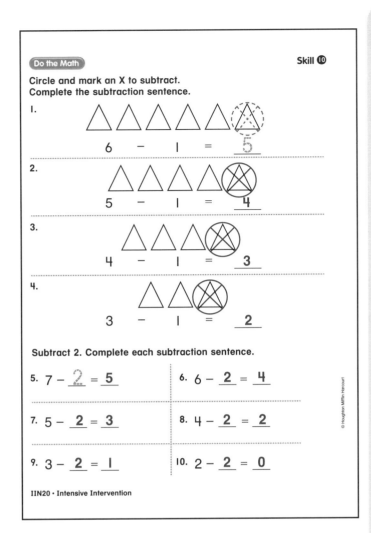

Do the Math

Skill ⑩

Circle and mark an X to subtract.
Complete the subtraction sentence.

1. $6 - 1 = 5$

2. $5 - 1 = 4$

3. $4 - 1 = 3$

4. $3 - 1 = 2$

Subtract 2. Complete each subtraction sentence.

5. $7 - 2 = 5$

6. $6 - 2 = 4$

7. $5 - 2 = 3$

8. $4 - 2 = 2$

9. $3 - 2 = 1$

10. $2 - 2 = 0$

IIN20 · Intensive Intervention

Do the Math

Have children look at student page **IIN20**.
Encourage children to model each problem.

Talk Math

- **What patterns do you see in Problems 1–4?**
  Possible answers: The number of triangles is
  one less each time. One triangle is subtracted
  each time. Each difference is one less than the
  difference before.
- **What patterns do you see in Problems 5–10?**
  Possible answers: The number you subtract
  from is one less each time. Two is subtracted
  each time. Each difference is one less than the
  difference before.

Check

Say: **Look back at Problems 1–4.** Ask: **If you
continued the pattern, what number sentence
would come next?** $2 - 1 = 1$

Alternative Teaching Strategy

Subtraction Pattern Stories

Objective To create story problems with
subtraction patterns

- Explain to children that they are going to help
  you tell subtraction stories about subtracting
  one.
- Start a story such as the following: **Maria likes
  to pick flowers for her friends. She picks 10
  flowers. She gives 1 flower to Sam.**
- Ask a volunteer to record on the board the
  subtraction sentence that shows how many
  flowers Maria has left. Say: **10 – 1 = 9.**
- Continue the story following a pattern of
  subtracting 1. Have children insert the missing
  numbers. **Maria now has _9_ flowers. She gives
  _1_ flower to Carson.**
- Ask a volunteer to record the subtraction
  sentence to show how many flowers Maria has
  left. Say: **9 – 1 = 8.**
- Repeat the process through 1 – 1 = 0. You might
  vary the activity by using the names of children
  in the class, or by having children act out the
  story using classroom objects.
- If time allows, tell stories about subtracting two.

Objective
To use a number line to count on to find sums to 12

Materials
number lines to 12

Vocabulary
count on
addend

Pre-Assess
Give children number lines and draw a number line labeled 0–12 on the board. Write 5 + 2 = ___ on the board. Tell children they will use a number line to find the sum. Ask: **On which number on the number line will you start to count on?** 5 **Which direction will you move to count on, left or right?** right **How many jumps will you make?** 2 jumps **What is the sum of 5 and 2?** 7

Common Misconception
• Children may move in the wrong direction on the number line to count on.

• To correct this, remind children that they are adding. Point out that numbers increase as you move to the right on the number line. So, to count on, always move to the right.

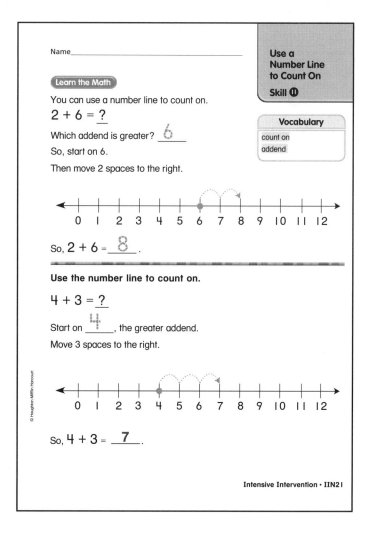

Learn the Math

Guide children through the first problem on student page **IIN21**. Have children use the number line to help them count on.

Talk Math

• **Why do you start at 6 on the number line to add 6 and 2?** because 6 is the greater addend

• **How do you know to make 2 jumps?** because you are counting on 2

• **Find 6. Now make 2 jumps to the right.** Have children trace over the arrows to make 2 jumps. **What is 2 + 6?** 8

Tell children that when they use a number line to count on, they will always make jumps to the right.

Guide children through the second problem by asking similar questions.

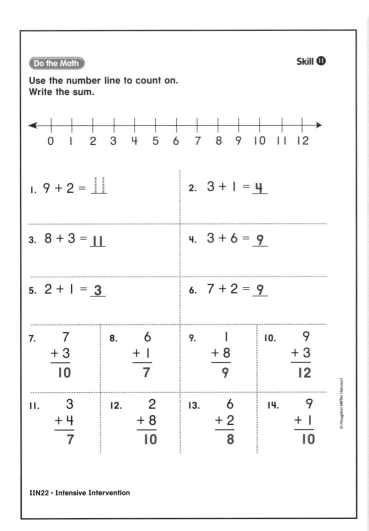

Do the Math

Use the number line to count on.
Write the sum.

Skill ⑪

0 1 2 3 4 5 6 7 8 9 10 11 12

1. $9 + 2 = \underline{11}$

2. $3 + 1 = \underline{4}$

3. $8 + 3 = \underline{11}$

4. $3 + 6 = \underline{9}$

5. $2 + 1 = \underline{3}$

6. $7 + 2 = \underline{9}$

| 7. $\begin{array}{r} 7 \\ +3 \\ \hline 10 \end{array}$ | 8. $\begin{array}{r} 6 \\ +1 \\ \hline 7 \end{array}$ | 9. $\begin{array}{r} 1 \\ +8 \\ \hline 9 \end{array}$ | 10. $\begin{array}{r} 9 \\ +3 \\ \hline 12 \end{array}$ |
| 11. $\begin{array}{r} 3 \\ +4 \\ \hline 7 \end{array}$ | 12. $\begin{array}{r} 2 \\ +8 \\ \hline 10 \end{array}$ | 13. $\begin{array}{r} 6 \\ +2 \\ \hline 8 \end{array}$ | 14. $\begin{array}{r} 9 \\ +1 \\ \hline 10 \end{array}$ |

IIN22 · Intensive Intervention

Alternative Teaching Strategy

Number Line Hop

Objective To use a number line to count on 1, 2, or 3 to find sums to 12

Materials adhesive tape, paper

• Make a large floor number line labeled 0–12 by taping several sheets of paper together.

• Tell children that they will use the number line to count on to add. Explain that they will stand next to the greater addend on the floor number line and then step to the right to count on.

• Invite a volunteer to use the floor number line. Tell children they will add 5 and 3. Ask: **On which number should [child's name] start?** 5 **How many spaces should [child's name] step?** 3 spaces Tell the child to step 3 spaces to the right on the number line. Ask: **What is the sum?** 8

• Continue the activity with different addition problems, adding 1, 2, or 3 to show sums to 12. Continue until each child has had the opportunity to use the floor number line.

Do the Math

Have children look at student page **IIN22**. Explain to children that they will use the number line at the top of the page for each problem. Encourage them to model the jumps on the number line as you work through each problem with them.

Talk Math

• **Look at the number sentence for Problem 1. On which number do you start?** You start on the greater addend, 9.

• **How do you know how many jumps to make on the number line?** You make as many jumps as the number you are counting on.

Check

Ask: **Which way do you make jumps on a number line to count on?** You make jumps to the right to count on.

Add Doubles
Skill ⑫

Objective
To use doubles facts to find sums

Materials
connecting cubes

Vocabulary
doubles fact
addend

Pre-Assess
Show children 2 connecting cubes in one hand and 2 connecting cubes in the other hand. Ask: **What doubles fact does this show?** 2 + 2 = 4
Show children 4 connecting cubes in one hand and 4 connecting cubes in the other hand. Ask: **What doubles fact does this show?** 4 + 4 = 8

Common Misconception
• Children may incorrectly find an odd number as the sum when using doubles.

• To correct this, have children use connecting cubes or other manipulatives to model each problem. You may also wish to point out that for every doubles fact, the sum is always an even number.

Learn the Math
Guide children through the first problem on student page **IIN23**. You might have children use connecting cubes to model each problem.

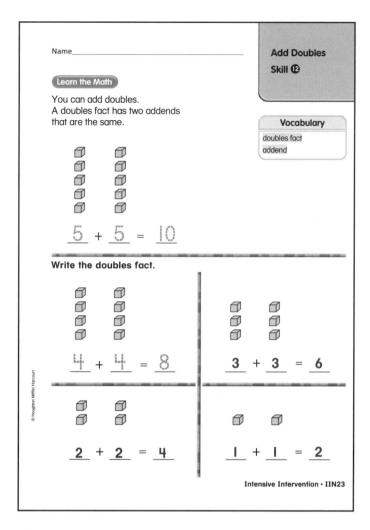

Talk Math

• **What number is doubled?** 5 Point out the two groups of 5 cubes. Have children circle each group.

• **How many cubes are there in all?** 10 cubes

• **What is the doubles fact?** 5 + 5 = 10 Have children trace the doubles fact.

Work with children to write the doubles facts in the remaining problems. You might have children use connecting cubes to model and record other doubles facts.

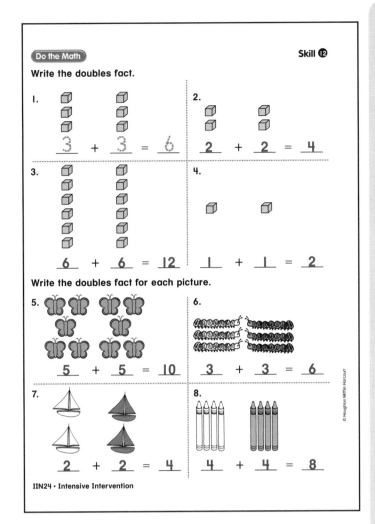

Do the Math

Have children look at student page **IIN24**. Point out to children that they will use the model or picture to write each doubles fact.

Talk Math

• **How will you find out which number to double?**
Count the cubes in each group.

• **What doubles fact is shown in Problem 2?**
2 + 2 = 4

Check

Ask: **Sam has 4 red marbles. Elise has 4 blue marbles. What doubles fact can you use to find how many marbles they have together?** 4 + 4 = 8

Alternative Teaching Strategy

Double the Number

Objective To use counters to add doubles

Materials two-color counters

• Give each pair of children 12 two-color counters.

• Explain that they will use the counters to add doubles.

• Have one child place one red counter on the table between them. Ask: **What number will you double?** 1 Have the other child show the double by placing a yellow counter on the table. **How many counters are there in all?** 2 Write ___ + ___ = ___ on the board. What is the doubles fact? 1 + 1 = 2

• Continue the activity with 2, 3, 4, 5, and 6 counters. After you have completed the activity with doubles facts up to 6 + 6 = 12, have children repeat the activity on their own and make a list of doubles facts from 1 + 1 = 2 through 6 + 6 = 12.

Doubles Plus One
Skill ⑬

Objective
To use doubles and doubles plus one to find sums to 12

Materials
connecting cubes

Vocabulary
doubles plus one

Pre-Assess
Show children 4 connecting cubes in one hand and 5 connecting cubes in the other hand. Ask: **What doubles fact can you use to help you find 4 + 5?** 4 + 4 = 8 Ask: **What is the sum of 4 and 5?** 4 + 5 = 9

Common Misconception
• When a greater number appears first in the problem, children may double that number and add one more.

• To correct this, remind children to always use the double of the lesser addend. Doubling the greater addend and subtracting one can be introduced later on.

Learn the Math

Guide children through the first problem on student page **IIN25**. Have children use connecting cubes to model the problem.

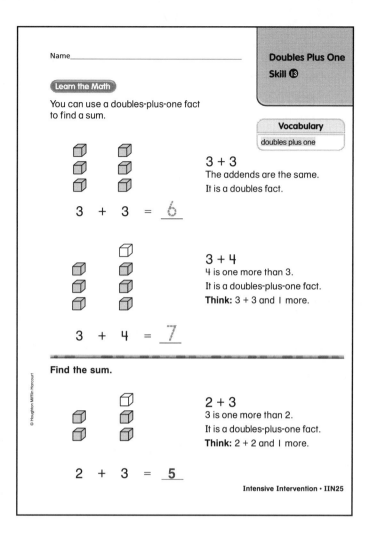

Learn the Math

You can use a doubles-plus-one fact to find a sum.

Vocabulary
doubles plus one

3 + 3
The addends are the same.
It is a doubles fact.

3 + 3 = 6

3 + 4
4 is one more than 3.
It is a doubles-plus-one fact.
Think: 3 + 3 and 1 more.

3 + 4 = 7

Find the sum.

2 + 3
3 is one more than 2.
It is a doubles-plus-one fact.
Think: 2 + 2 and 1 more.

2 + 3 = 5

Intensive Intervention • IIN25

Talk Math

• **What is the doubles fact?** 3 + 3 = 6

• **How many cubes are added to 6 to find 3 + 4?** 1

• **What is 3 + 4?** 7

• **Why is 3 + 4 = 7 a doubles-plus-one fact?** You can use the doubles fact 3 + 3 = 6 and then add 1 to find 3 + 4 = 7.

Help children work through the second problem.

Do the Math

Use ⬜ . Write the addition sentence.

1. $4 + 5 = 9$

2. $2 + 1 = 3$

3. $5 + 6 = 11$

4. $4 + 3 = 7$

5. $3 + 2 = 5$

6. $5 + 4 = 9$

7. $1 + 2 = 3$

8. $6 + 5 = 11$

IIN26 · Intensive Intervention

© Houghton Mifflin Harcourt

Do the Math

Have children look at student page **IIN26**. Help children understand that the white cube in each problem shows 1 added to the double. Encourage children to use connecting cubes to model each problem.

Talk Math

- **Which addend will you double—the greater or lesser addend?** the lesser addend

- **What will you do to find the sum of the doubles-plus-one fact after you find the doubles sum?** add 1

Check

Ask: **To find 4 + 5, how could you use 4 + 4?** I find 4 + 4 and then add 1 to it.

Alternative Teaching Strategy

Double the Number

Objective To use a number line to find doubles plus one sums

- Draw a number line on the board labeled 0–12. Write 4 + 4 = ___ under the number line. Explain to children that they can use the number line to find the sum.

- Ask: **What number should you start on?** 4 Place your finger on 4. Ask: **How many jumps should you make to add 4?** 4 **Which direction should you move?** to the right Make 4 jumps to the right. Ask: **What is 4 + 4?** 8 Complete the number sentence under the number line.

- Explain that you can also use a number line to find doubles-plus-one sums. Write 4 + 5 = ___ under the number line. Say: **We already know that 4 + 4 = 8, how many more jumps do we need to make to find 4 + 5?** 1 Make one jump to the right from 8 to 9. Ask: **What is 4 + 5?** 9 Complete the number sentence under the number line.

- Write 3 + 4 = ___ on the board. Ask: **Which number do you want to double?** 3 Ask a volunteer to circle 3 on the number line.

- Ask: **What is 3 + 3?** 6 Have another volunteer circle 6 in a different color on the number line. Ask: **What is 3 + 4?** 7 Have a third volunteer circle 7 in a different color.

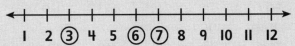

- Repeat the process using other doubles-plus-one facts.

© Houghton Mifflin Harcourt

Add in Any Order
Skill ⑭

Objective
To add using the Order Property

Materials
two-color counters

Vocabulary
addend
order

Pre-Assess
Write 2 + 3 = ___ on the board. Ask: **What is 2 + 3?** 5 Write 5 in the addition sentence. Then write 3 + 2 = ___ on the board. Ask: **What is 3 + 2?** 5 Write 5 in the addition sentence. **How is 2 + 3 = 5 the same as 3 + 2 = 5? How is it different?** The addends and the sum are the same but the order of the addends is different.

Common Misconception

- Children may have difficulty understanding that the order of addends may be changed in an addition sentence.

- To correct this, bring 5 children to the front of the classroom. Arrange them in a group of 2 and a group of 3. Begin with the group of 2 and count the children from I to 5. Do the same thing again, but begin with the group of 3. Stress that there will be 5 children in all, no matter how you count them.

Learn the Math

Guide children through the first problem on student page **IIN27**. Have children use two-color counters to model the problem.

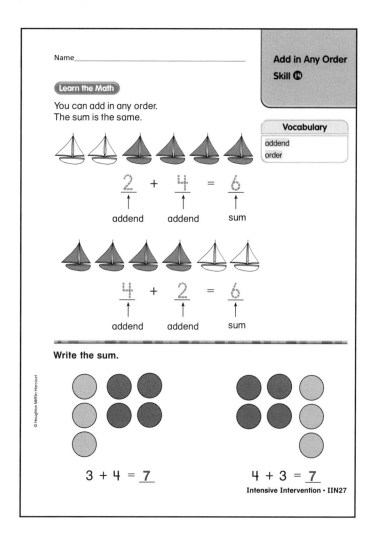

Learn the Math

You can add in any order.
The sum is the same.

Vocabulary
addend
order

2 + 4 = 6
↑   ↑   ↑
addend addend sum

4 + 2 = 6
↑   ↑   ↑
addend addend sum

Write the sum.

3 + 4 = 7   4 + 3 = 7

Intensive Intervention · IIN27

Talk Math

- **How many white sailboats are there in the first picture?** 2 **In the second picture?** 2

- **How many gray sailboats are there in the first picture?** 4 **In the second picture?** 4

- **Count the sailboats. How many are there in all in the first picture?** 6 **In the second picture?** 6

- **Does changing the order of the addends change the sum?** no

Have children trace over the addends and sums in the first problem. Repeat the questions as children work through the second problem.

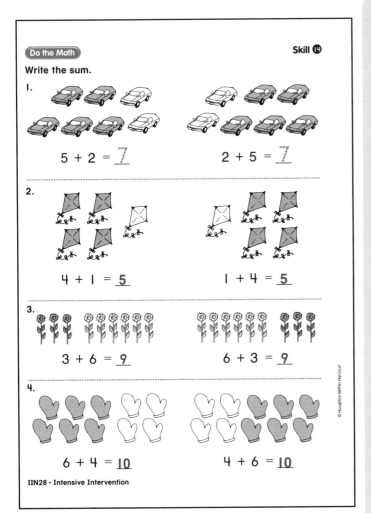

Do the Math

Write the sum.

1.

$5 + 2 = 7$ $2 + 5 = 7$

2.

$4 + 1 = \underline{5}$ $1 + 4 = \underline{5}$

3.

$3 + 6 = \underline{9}$ $6 + 3 = \underline{9}$

4.

$6 + 4 = \underline{10}$ $4 + 6 = \underline{10}$

IIN28 • Intensive Intervention

Alternative Teaching Strategy

Surprise Sentences

Objective To add using the Order Property

Materials two-color counters

- Write several addition sentences with sums to 12, such as $4 + 7 =$ ___, on small pieces of paper. Fold the pieces of paper and place them in a hat or other open container. Have a child take an addition sentence out of the hat and read it aloud. Write the addition sentence on the board.

- Give each child 12 two-color counters. Have them use the counters to find the sum.

- Reverse the addends. For example, if the first sentence is $3 + 6 = 9$, write $6 + 3 =$ ___ next to it on the board.

- Have children find the sum. Ask: **Is the sum the same? Why or why not?** Yes, it is the same because you can change the order of the addends and the sum stays the same.

- Have another child take another addition sentence out of the hat. Repeat the activity until all of the addition sentences have been completed.

Do the Math

Have children look at student page **IIN28**. Help children see that for each problem there are two addition sentences. Each sentence has the addends in a different order. Encourage children to use two-color counters to model each problem.

Talk Math

- **Why are the sums the same in each problem?** because it does not matter in which order you add the addends, the sum will be the same

Check

Ask: **The sum of 5 and 6 is 11. What is the sum of 6 and 5?** 11 **How do you know?** It is the same as the sum of 5 and 6 because it does not matter in which order you add the addends.

Objective
To add tens using base-ten blocks

Materials
base-ten blocks

Pre-Assess

Model 30 and 40 with base-ten blocks. Point to the model for 30. Ask: **What number does the model show?** 30 Point to the model for 40. **What number does the model show?** 40 Write 30 + 40 = __ on the board. Say: **Look at the models to help you.** Join the tens. Ask: **What is 30 + 40?** 70 Repeat with other addition problems such as 20 + 20, 50 + 30, and 10 + 60.

Common Misconception

• Children may incorrectly name the value of the digit in the tens place as ones. For example, a child may give the value for the digit 2 in 25 as 2 ones, not 2 tens.

• To correct this, have children add with base-ten blocks to see that the value of the digit in the tens place is a number of tens, not a number of ones.

Learn the Math

Have children use base-ten blocks to model the first problem on student page **IIN29** as you guide them through the problem.

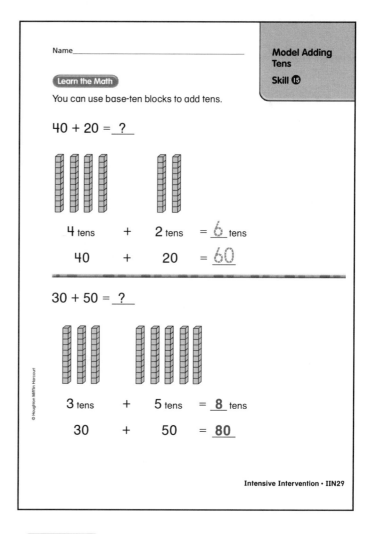

Learn the Math

You can use base-ten blocks to add tens.

$40 + 20 = \underline{?}$

4 tens + 2 tens = $\underline{6}$ tens

40 + 20 = $\underline{60}$

$30 + 50 = \underline{?}$

3 tens + 5 tens = $\underline{8}$ tens

30 + 50 = $\underline{80}$

Intensive Intervention · IIN29

Talk Math

• **How many tens are in 40?** 4 tens

• **How many tens are in 20?** 2 tens

• **What is 4 tens + 2 tens?** 6 tens

• **How many ones are equal to 6 tens?** 60 ones

• **What is 40 + 20?** 60

Have children trace the number of tens and the sum. Repeat the questions as you guide children through the second problem.

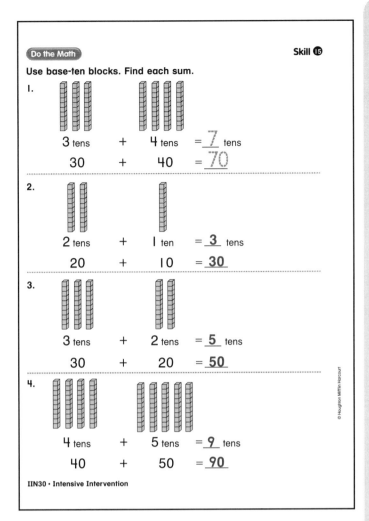

Use base-ten blocks. Find each sum.

1.

3 tens   +   4 tens   = _7_ tens

30   +   40   = _70_

2.

2 tens   +   1 ten   = _3_ tens

20   +   10   = _30_

3.

3 tens   +   2 tens   = _5_ tens

30   +   20   = _50_

4.

4 tens   +   5 tens   = _9_ tens

40   +   50   = _90_

IIN30 · Intensive Intervention

© Houghton Mifflin Harcourt

Do the Math

Have children look at student page **IIN30**. Encourage children to use tens blocks to model each problem.

Talk Math

- **What does each tens block stand for?** 10 ones

- **How do you use the models to help you add tens?** Count the number of tens blocks to find how many tens.

Check

Show children base-ten blocks for 20 and base-ten blocks for 40. Ask: **How many tens are there in all?** 6 tens **How much is 6 tens?** 60 **What is 20 + 40?** 60

Alternative Teaching Strategy

Ten More

Objective To practice adding tens using base-ten blocks

Materials base-ten blocks

- Give each pair of children 10 tens blocks. Ask the children to each take 5 of the tens blocks.

- Write 10 + 10 = ___ on the board. Say: **Each of you put one of your tens blocks in the middle. Use them to find 10 + 10.** Ask: **What is 10 + 10?** 20

- Write 20 as the sum in the addition sentence on the board. Below the first addition sentence, write 20 + 10 = ___ . Say: **Leave the 2 tens blocks in front of you. Now add 1 more ten. Use your models to find 20 + 10.** Ask: **What is 20 + 10?** 30 Write 30 as the sum in the addition sentence on the board.

- Next, write 30 + 10 = ___ on the board. Say: **Add 1 more tens block to the 3 tens blocks that are already on the table.** Ask: **What is 30 + 10?** 40 Write 40 as the sum on the board.

- Repeat the activity until children have added 90 + 10. You may wish to alter the activity by adding 20, 30, or 40 at a time.

Objective
To use a number line to count back

Vocabulary
count back
number line

Pre-Assess

Draw a number line labeled 0–12 on the board. Then write 6 – 2 = ___ on the board. Tell children they can use a number line to count back. Ask: **Will you count on or count back to subtract?** count back **On which number will you start to count back on the number line?** 6 **Which way will you move on the number line?** left **How many jumps will you make?** 2 jumps **What is 6 – 2?** 4

Common Misconception

• Children may move in the wrong direction on the number line to count back.

• To correct this, remind children that they are subtracting. Point out that numbers decrease as you move to the left on the number line. So, to count back, always move to the left.

Learn the Math

Guide children through the first problem on student page **IIN31**. Have children use the number line to help them count back.

Talk Math

• **On which number do you start?** You start on the number you will subtract from.

• **Why do you make jumps to the left on the number line to count back?** You make jumps to the left because you are subtracting and the numbers decrease as you move to the left.

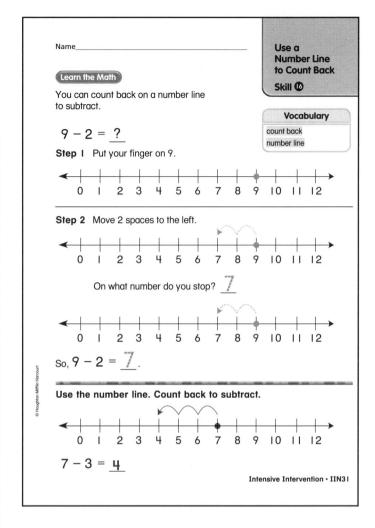

• **How do you know how many jumps to make?** The number of jumps is the number you are subtracting.

• Have children put a finger on 9, and count back two spaces. **Where do you stop?** 7 **What is 9 – 2?** 7

Work through the second problem with children. Have them draw the arrows to show the jumps on the number line. Then have them write the difference.

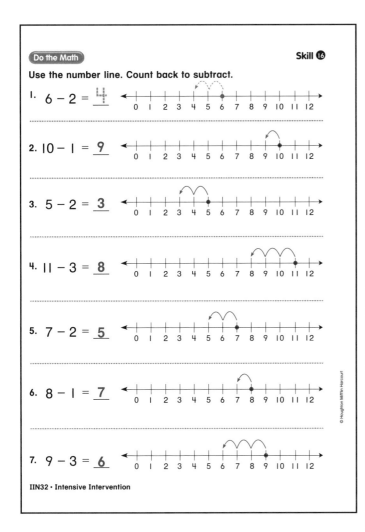

Use the number line. Count back to subtract.

Skill **16**

1. $6 - 2 = \underline{4}$

2. $10 - 1 = \underline{9}$

3. $5 - 2 = \underline{3}$

4. $11 - 3 = \underline{8}$

5. $7 - 2 = \underline{5}$

6. $8 - 1 = \underline{7}$

7. $9 - 3 = \underline{6}$

IIN32 · Intensive Intervention

© Houghton Mifflin Harcourt

Do the Math

Have children look at student page **IIN32**. Point out that they will use a number line for each problem. Direct them to draw the jumps on the number line as you guide them through each problem.

Talk Math

• **Which direction will you make jumps on the number line to count back?** to the left

• **How will you know how many jumps to make?** Make as many jumps as the number you are subtracting.

Check

Ask: **Anna subtracts 2 from 5 and gets a difference of 7. Is she correct?** no **What mistake could she have made?** She moved to the right on the number line when she needed to go left.

Alternative Teaching Strategy

Go Back

Objective To use a number line to count back

• Draw a number line labeled 0–12 on the board. Ask a child to choose a number between 4 and 12. Write the number on the board. Have a second child choose 1, 2, or 3. Use the two numbers to write a subtraction sentence such as $8 - 2 = \underline{}$.

• Tell children they will practice using the number line to count back. Tell them that counting back is a good strategy to use when subtracting 1, 2, or 3 from a number.

• Ask: **On which number should you start?** You start on the first number in the subtraction sentence. **Which direction will you make jumps to count back?** to the left **How many jumps will you make?** You make the same number of jumps as the second number in the subtraction sentence.

• Have a volunteer place his or her finger on the first number. Then have the child count the number of jumps to the left to find the difference. Ask the child to draw a square around the first number and a circle around the difference on the number line.

• Continue the activity to allow children to practice counting back on a number line. Give all children a chance to suggest numbers and to count back.

Related Subtraction Facts to 12
Skill ⑰

Objective
To identify related subtraction facts to 12

Materials
counters

Vocabulary
related subtraction fact

Pre-Assess

Write 9 − 6 = ___ on the board. Ask: **What is the difference?** 3 Then write 9 − 3 = ___ on the board. **What is the difference?** 6 **What do you notice about these two subtraction facts?** They both use the numbers 3, 6, and 9. Write 11 − 7 = 4 and 11 − 4 = ___ on the board. **What is 11 − 4?** 7 **How do you know?** If 11 − 7 = 4, then 11 − 4 = 7. **Are these two subtraction facts related? Explain.** Yes, they both use the same numbers.

Common Misconception

- Children may think that for two subtraction facts to be related, all of the numbers in the two facts must be rearranged, instead of always subtracting from the same number.

- To correct this, point out that in related subtraction facts, the number they subtract from will always stay the same. The positions of the other two numbers will change.

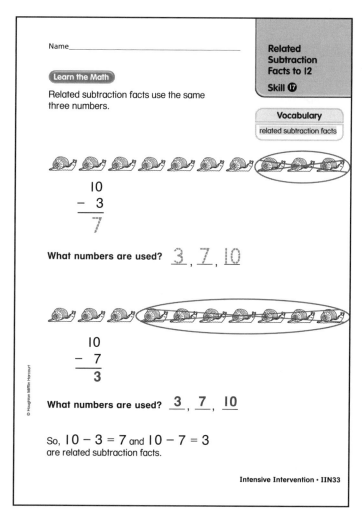

Learn the Math

Related subtraction facts use the same three numbers.

Vocabulary
related subtraction facts

$$\begin{array}{r} 10 \\ -\ 3 \\ \hline 7 \end{array}$$

What numbers are used? 3, 7, 10

$$\begin{array}{r} 10 \\ -\ 7 \\ \hline 3 \end{array}$$

What numbers are used? 3, 7, 10

So, 10 − 3 = 7 and 10 − 7 = 3 are related subtraction facts.

Intensive Intervention · IIN33

Learn the Math

Guide children through the problem on student page **IIN33**. Have children trace the difference in the first subtraction problem.

Talk Math

- **What is 10 − 3?** 7

- **What is 10 − 7?** 3

- **What three numbers are used in 10 − 3 = 7 and 10 − 7 = 3?** 3, 7, and 10

- **Which number do you subtract from in both subtraction facts?** 10

- **How do you know that 10 − 3 = 7 and 10 − 7 = 3 are related subtraction facts?** because they have the same three numbers

Have children write the difference to 10 − 7 and then write the three numbers that are used in the problem.

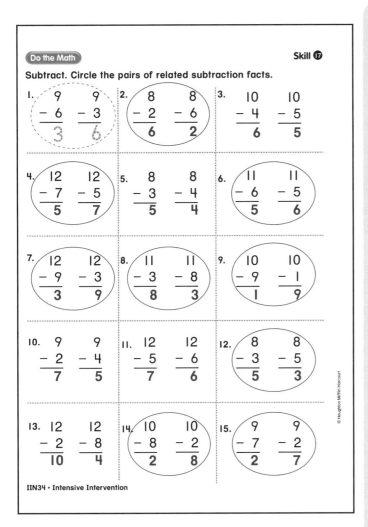

Alternative Teaching Strategy

Are They Related?

Objective To find related subtraction facts to 12 and to use subtraction facts to write related subtraction facts

Materials counters

• Give each pair of children 12 counters. Write 11 – 5 on the board. Tell children to use their counters to find the difference by showing 11 counters and then pushing a group of 5 to the side. Ask: **What is 11 – 5?** 6

• Point out that the 11 counters have been separated into a group of 5 and a group of 6.

• Tell children to use their counters to find 11 – 6 without moving the counters. Ask: **What is 11 – 6?** 5

• Write 11 – 5 = 6 and 11 – 6 = 5 on the board. Ask: **Are these subtraction facts related? Why?** Yes, they both use the same three numbers.

• Give children other subtraction facts to model with their counters. Each time have them model and record the subtraction fact and its related subtraction fact. Discuss why they are related.

Do the Math

Have children look at student page **IIN34**. Explain to children that they will first subtract to find the difference in each problem. Then they will compare the two problems in each pair to see if they are related. Remind them that related subtraction facts use the same three numbers.

Talk Math

• **What is 9 – 6?** 3 **What is 9 – 3?** 6

• **Are they related subtraction facts? Explain.** Yes, they both use the same three numbers.

Check

Ask: **Are 11 – 6 = 5 and 11 – 4 = 7 related subtraction facts? Explain.** No, they do not have the same three numbers. Only 11 is the same in both facts.

Fact Families to 12
Skill ⑱

Objective
To identify fact families

Materials
connecting cubes

Vocabulary
fact family

Pre-Assess

Write the following on the board:

9 + 3 = ___   12 − 9 = ___
3 + 9 = ___   12 − 3 = ___

Ask: **What is 9 + 3?** 12 **What is 3 + 9?** 12 **What is 12 − 9?** 3 **What is 12 − 3?** 9 **How are these facts related?** They all use the numbers 3, 9, and 12. **What numbers are in the fact family?** 3, 9, and 12

Common Misconception

- Children may try to use a number that is not included in the fact family in one or more of their number sentences.

- To correct this, point out that there are three numbers used in a fact family, unless the fact is a double. They should use the three numbers to write two addition sentences and two subtraction sentences. Have them check their addition and subtraction sentences to make sure they have used only the same three numbers in each sentence.

Learn the Math

Give children 2 connecting cubes of one color and 3 connecting cubes of another color. Have them use the connecting cubes to model the problem on student page **IIN35**.

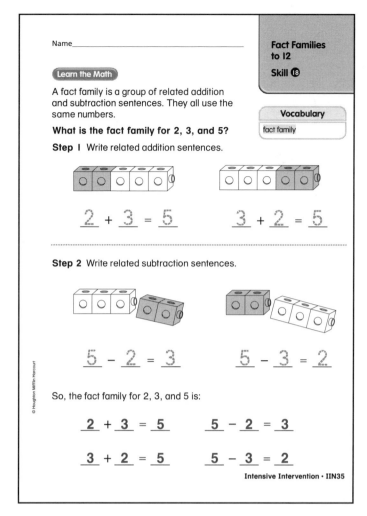

Learn the Math

A fact family is a group of related addition and subtraction sentences. They all use the same numbers.

What is the fact family for 2, 3, and 5?

Step 1 Write related addition sentences.

2 + 3 = 5   3 + 2 = 5

Step 2 Write related subtraction sentences.

5 − 2 = 3   5 − 3 = 2

So, the fact family for 2, 3, and 5 is:

2 + 3 = 5   5 − 2 = 3

3 + 2 = 5   5 − 3 = 2

Intensive Intervention · IIN35

Talk Math

- **Use your connecting cubes to model 2 + 3. What is 2 + 3?** 5

- **Use the same connecting cubes. What is 3 + 2?** 5 Have children trace the related addition sentences.

- **Connect all of the connecting cubes. Take away 2. What is 5 − 2?** 3

- **Put the connecting cubes back together. Take away 3. What is 5 − 3?** 2 Have children trace the related subtraction sentences.

- **How are the addition and subtraction sentences the same?** They all use the numbers 2, 3, and 5.

Have children write the addition and subtraction sentences for the fact family.

Do the Math

Add or subtract to complete the fact family.
Write the numbers in the fact family.

1. $6 + 3 = \underline{9}$ $9 - 6 = \underline{3}$ $\boxed{3}, \boxed{6}, \boxed{9}$
   $3 + 6 = \underline{9}$ $9 - 3 = \underline{6}$

2. $5 + 2 = \underline{7}$ $7 - 5 = \underline{2}$ $\boxed{2}, \boxed{5}, \boxed{7}$
   $2 + 5 = \underline{7}$ $7 - 2 = \underline{5}$

3. $1 + 4 = \underline{5}$ $5 - 1 = \underline{4}$ $\boxed{1}, \boxed{4}, \boxed{5}$
   $4 + 1 = \underline{5}$ $5 - 4 = \underline{1}$

4. $5 + 5 = \underline{10}$ $10 - 5 = \underline{5}$ $\boxed{5}, \boxed{10}$

Write the number sentences to make a fact family.

5. 4, 5, 9

   $\boxed{4} + \boxed{5} = \boxed{9}$ $\boxed{9} - \boxed{4} = \boxed{5}$

   $\boxed{5} + \boxed{4} = \boxed{9}$ $\boxed{9} - \boxed{5} = \boxed{4}$

6. 3, 7, 10

   $\boxed{3} + \boxed{7} = \boxed{10}$ $\boxed{10} - \boxed{3} = \boxed{7}$

   $\boxed{7} + \boxed{3} = \boxed{10}$ $\boxed{10} - \boxed{7} = \boxed{3}$

IIN36 · Intensive Intervention

© Houghton Mifflin Harcourt

Do the Math

Point out that in Problems 1–4 on student page **IIN36**, students need to add or subtract and then write the numbers in the fact family. Point out that in Problem 4 there are only two number sentences in the fact family. Explain that since there are only two different numbers, you can only make one addition sentence and one subtraction sentence.

In Problems 5 and 6, students need to use the three numbers to write the sentences in the fact family.

Talk Math

• **How do you know that addition and subtraction sentences in a fact family are related?** They use the same numbers.

• **How are the number sentences in a fact family different?** The order of the numbers in each number sentence is different.

Check

Ask: **When does a fact family only have two number sentences? Explain.** A fact family has only two sentences when two of the numbers are the same, such as 3 + 3 = 6 and 6 − 3 = 3.

Alternative Teaching Strategy

Find the Fact Family

Objective To write fact families to 12

Materials number cubes

• Have two children each toss a number cube. Write each of the numbers tossed on the board. Ask: **What is [first number] + [second number]?** Check children's answers.

• Tell children they can write a fact family for these three numbers. Explain that they already made one addition sentence when they added the first two numbers. Have them write this sentence.

• Point to the first addition sentence. Ask: **What do you do to the addends to write a second addition sentence?** I change the places of the addends. **So, what is the second addition sentence?** Check children's answers. Have them write the sentence.

• Say: **Next, we will write the two subtraction sentences.** Ask: **Which of the three numbers will you subtract from?** the greatest number

• Say: **Write a subtraction sentence in which you subtract one of the other numbers from the greatest number.**

• Ask: **What will you do to write the second subtraction sentence?** Change the places of the two other numbers. Ask: **What is the second subtraction sentence?** Check children's answers. Have them write the sentence.

• Give pairs of children number cubes. Have them repeat the activity on their own.

Intensive Intervention • IIN36

Think Addition to Subtract
Skill ⑲

Objective
To use addition as a strategy to subtract from 12 or less

Materials
connecting cubes

Pre-Assess
Write the following on the board:

$$8 + 3 = 11 \qquad 11 - 3 = \underline{\quad}$$

Say: **Use the addition fact 8 + 3 = 11 to help you find 11 – 3.** Ask: **What is 11 – 3?** 8 **How does knowing that 8 + 3 = 11 help you find 11 – 3?** The same numbers are used in both sentences. If 8 + 3 = 11, then 11 – 3 has to equal 8 because they are related facts.

Common Misconception

• Children may use the wrong addition sentence when they think addition to subtract.

• To correct this, point out that related addition and subtraction facts use the same three numbers, only in a different order. Guide children to see that each number is used once in each fact, unless the fact is a double.

Learn the Math

Give children 6 connecting cubes of one color and 2 connecting cubes of another color. Have them use the connecting cubes to model the first problem on student page **IIN37**.

Name_____

Learn the Math

Think Addition to Subtract
Skill ⑲

You can use addition facts to help you subtract.

$8 - 6 = \underline{?}$

Think: $6 + \underline{2} = 8$

Since $6 + \underline{2} = 8$, then $8 - 6 = \underline{2}$.

$10 - 3 = \underline{?}$

Think: $3 + \underline{7} = 10$

Since $3 + \underline{7} = 10$, then $10 - 3 = \underline{7}$.

$9 - 6 = \underline{?}$

Think: $6 + \underline{3} = 9$

Since $6 + \underline{3} = 9$, then $9 - 6 = \underline{3}$.

Intensive Intervention • IIN37

Talk Math

• **Use your connecting cubes to model 6 + 2. What is 6 + 2?** 8

• **Now show 8 – 6 with connecting cubes. How many are left?** 2 cubes **So, what is 8 – 6?** 2

• **Look at the addition fact and the subtraction fact. How are they alike?** They both use the same three numbers.

• **If you know that 6 + 2 = 8, how can you find 8 – 6?** Possible answer: If 6 + 2 = 8, then 8 – 6 = 2 because they are related facts.

Have children trace the numbers in the first problem. Then work through the second and third problems with them. Have them write the missing addends and the differences.

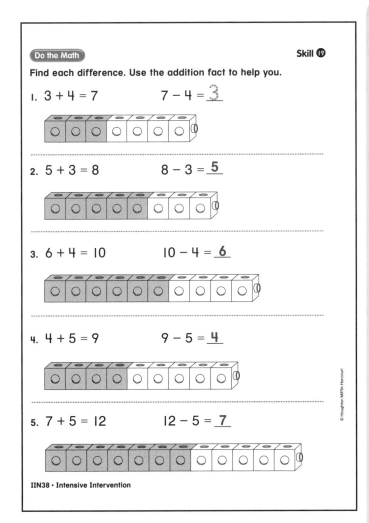

Find each difference. Use the addition fact to help you.

1. 3 + 4 = 7   7 – 4 = _3_

2. 5 + 3 = 8   8 – 3 = _5_

3. 6 + 4 = 10   10 – 4 = _6_

4. 4 + 5 = 9   9 – 5 = _4_

5. 7 + 5 = 12   12 – 5 = _7_

IIN38 · Intensive Intervention

© Houghton Mifflin Harcourt

Alternative Teaching Strategy

Add to Subtract

Objective To use a number line to explore using addition to subtract

- Draw a number line labeled 0–12 on the board. Then write 7 – 5 = ___ and 5 + ___ = 7 under the number line. Explain that instead of beginning on the number line at the number they are subtracting from and moving left, they will begin at the number that is being subtracted and move right.

- Place your finger on 5 on the number line. Say: **Now I will move to 7 on the number line. Count the jumps with me…one…two. So, 5 + 2 = 7.** Complete the addition sentence on the board by writing 2. Ask: **Since 5 + 2 = 7, what is 7 – 5?** 2

- Continue the activity with different subtraction sentences. Then have children continue the activity in pairs. Challenge children to write three subtraction sentences and then have their partner find each difference by using the number line to add.

Do the Math

Point out to children that on student page **IIN38** they are asked to find the difference. The addition sentence is there to help them. Encourage them to use connecting cubes if they need to as you work through the problems together.

Talk Math

- **What is the difference in the subtraction sentence called in the related addition sentence?** an addend

- **What does it mean when an addition sentence and a subtraction sentence are related?** They use the same three numbers.

Check

Ask: **How does knowing 4 + 4 = 8, help you find 8 – 4?** When you add 4 to 4 and get 8, you know that when you take the 4 away you will still have the 4 you started with.

Model Subtracting Tens
Skill 20

Objective
To subtract tens using base-ten models

Materials
base-ten blocks

Pre-Assess
Write 60 – 20 on the board. Have students model 60 with base-ten blocks. Ask: **What number does each tens block show?** 10 **What number does the model show?** 60 **Look at the number sentence. How much should you take away?** 20 or 2 tens Have students model the subtraction sentence. **What is 60 – 20?** 40 Repeat with other subtraction problems such as 50 – 10, 70 – 60, and 30 – 10.

Common Misconception
- Children may add instead of subtract.
- To correct this, remind children to look carefully at the symbol in the number sentence. If it is a minus symbol, they should subtract the second number from the first number to find the difference.

Learn the Math

Have children use base-ten blocks to model the first problem on student page **IIN39** as you guide them through the problem.

Name_____

Model Subtracting Tens
Skill 20

Learn the Math

You can use base-ten blocks to subtract tens.

$50 - 20 = \underline{?}$

How many tens are there in all? 5 tens

How many tens are taken away? 2 tens

How many tens are left? 3 tens

$5 \text{ tens} - 2 \text{ tens} = \underline{3} \text{ tens}$
$50 - 20 = \underline{30}$

$60 - 40 = \underline{?}$

How many tens in are there all? 6 tens

How many tens are taken away? 4 tens

How many tens are left? 2 tens

$6 \text{ tens} - 4 \text{ tens} = \underline{2} \text{ tens}$
$60 - 40 = \underline{20}$

Intensive Intervention · IIN39

© Houghton Mifflin Harcourt

Talk Math

- **How many tens are in 50?** 5 tens
- **How many tens are in 20?** 2 tens
- **How many tens should you take away?** 2 tens
- **What is 5 tens – 2 tens?** 3 tens
- **How many ones are equal to 3 tens?** 30 ones
- **What is 50 – 20?** 30

Have children trace the number of tens and the difference. Repeat the questions as you guide children through the second problem. Have them write the number of tens and the difference.

© Houghton Mifflin Harcourt

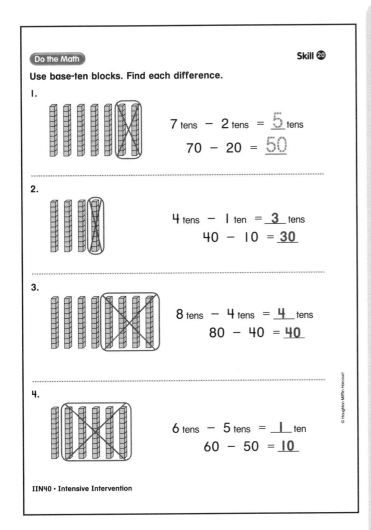

Do the Math   Skill **20**

Use base-ten blocks. Find each difference.

1.

7 tens − 2 tens = 5 tens

70 − 20 = 50

2.

4 tens − 1 ten = 3 tens

40 − 10 = 30

3.

8 tens − 4 tens = 4 tens

80 − 40 = 40

4.

6 tens − 5 tens = 1 ten

60 − 50 = 10

IIN40 · Intensive Intervention

© Houghton Mifflin Harcourt

Do the Math

Have children use base-ten blocks to model the problems on student page IIN40. Remind them that they should subtract the second number from the first number.

Talk Math

• **In Problem 1, how many tens do you subtract from 7 tens?** 2 tens

• **How do you use the models to help you subtract tens?** Possible answer: Take away the number of tens in the second number from the number of tens in the first number to find the difference.

Check

Ask: **How can you use base-ten blocks to subtract 40 from 90?** Show 9 tens then take away 4 tens. You will have 5 tens, or 50, left.

Alternative Teaching Strategy

Draw and Cross Out the Tens

Objective To draw pictures to subtract tens

Materials paper, crayons or pencils

• Have each child draw a picture of 8 tens blocks on a piece of paper.

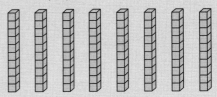

• Write 80 − 10 = ___ on the board. Ask: **How many tens will you subtract?** 1 ten Say: **Circle and cross out 1 ten.** Ask: **What is 80 − 10?** 70

• Write 70 as the difference in the subtraction sentence on the board. Below the first subtraction sentence, write 70 − 20 = ___. Ask: **How many tens will you subtract?** 2 tens **So, how many tens will you circle and cross out?** 2 tens Say: **Circle and cross out 2 tens.** Ask: **What is 70 − 20?** 50

• Write 50 as the difference in the subtraction sentence on the board. Next, write 50 − 30 = ___ on the board. Ask: **How many tens will you subtract?** 3 tens **So, how many tens will you circle and cross out?** 3 tens Say: **Circle and cross out 3 tens.** Ask: **What is 50 − 30?** 20

• You may wish to repeat the activity subtracting different numbers of tens.

Read a Tally Table
Skill ㉑

Objective
To read and interpret a tally table

Vocabulary
tally table
tally mark

Pre-Assess
Draw this tally table on the board.

Tu's Marbles	
green	⊬⊬ I
red	⊬⊬ II
white	IIII

Ask: **How many green marbles does Tu have?** 6 green marbles **How many red marbles does Tu have?** 7 red marbles **How many white marbles does Tu have?** 4 white marbles **Does Tu have more red or more white marbles?** more red marbles **How many marbles does Tu have in all?** 17 marbles

Common Misconception
• Children may have difficulty counting tally marks, particularly from 5 on.

• To correct this, practice counting tally marks with children. Show them a group of 5 tally marks and count the 5 tally marks with them. Explain that the fifth mark is always a diagonal slash through the first four marks. Then draw 8 tally marks and have them begin at 5 and count 5, 6, 7, 8. Emphasize that when they see groups of 5 tally marks, they can count by fives.

Learn the Math
Work through the questions about the tally table on student page IIN4I with children. Read each question aloud and help them answer it. Point to the information in the table that they can use to answer each question.

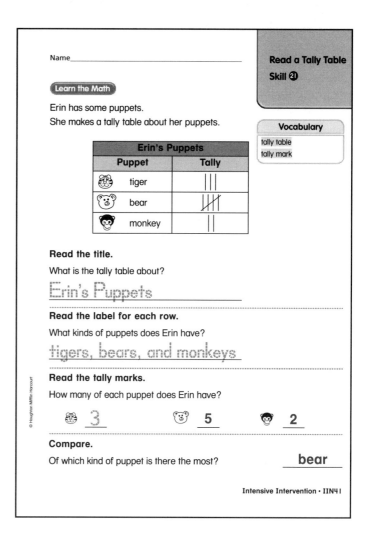

Talk Math

• **What information tells you what the table is about?** the title

• **How do you know what three kinds of puppets Erin has?** I can read the label for each row.

• **How do you find out how many of each kind of puppet Erin has?** I can count the tally marks for each kind of puppet.

• **How can you find out if Erin has more tiger puppets or more bear puppets?** I can compare the tally marks for tiger puppets and bear puppets.

Have children use the tally table to answer each question.

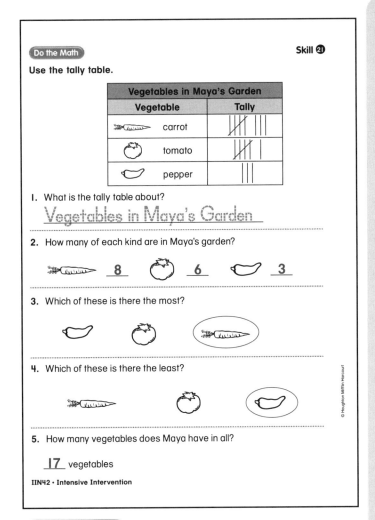

Use the tally table.

Vegetables in Maya's Garden	
Vegetable	**Tally**
carrot	//// ////
tomato	//// /
pepper	///

1. What is the tally table about?

   Vegetables in Maya's Garden

2. How many of each kind are in Maya's garden?

   8   6   3

3. Which of these is there the most?

4. Which of these is there the least?

5. How many vegetables does Maya have in all?

   17 vegetables

IIN42 • Intensive Intervention

Do the Math

Remind children they will use the information in the tally table on student page **IIN42** to answer the questions. Read the title and the labels for each row aloud. Then count the tally marks in each row with children.

Talk Math

- **How can you find how many of each kind of vegetable Maya has?** I can count and compare the tally marks.

- **Which of the vegetables is there the most? Explain.** There are more carrots than any other vegetable because carrots has the greatest number of tally marks.

Check

Ask: **How do you know if Maya has more tomatoes or more peppers in her garden? Explain.** I can compare the tally marks for tomatoes and peppers. There are 6 tally marks for tomato and 3 tally marks for pepper. So, Maya has more tomatoes than peppers.

Alternative Teaching Strategy

Tally Tables

Objective To read and interpret tally tables

- Make a tally table like the one shown below on the board.

Favorite Activities	
swimming	
hiking	
biking	

- Tell children to think about which of the three activities is their favorite. Say: **Raise your hand if swimming is your favorite activity.** Count the number of hands with children. Then make that number of tally marks next to swimming in the table. Have children count the tally marks with you. Repeat for hiking and biking.

- Once the tally table is completed, ask questions about the information. Ask: **How many children chose swimming?** Answers may vary. **Which activity did the most children choose?** Answers may vary. **Did more children choose hiking or biking?** Answers may vary.

- Continue with other questions about the tally table. You may also wish to create another tally table with different activities or a different topic.

Intensive Intervention • IIN42

Objective
To read and interpret a picture graph

Vocabulary
picture graph

Pre-Assess

Draw this picture graph on the board. Color each circle the appropriate color.

Our Favorite Colors							
blue	O	O	O	O	O	O	O
green	O	O	O				
red	O	O	O	O	O		

Ask: **How many children chose blue?** 7 children
How many children chose green? 3 children
How many children chose red? 5 children **Which color did the most children choose?** blue

Common Misconception

• Children may not understand that each picture stands for one object.

• To correct this, explain that each picture stands for one. Go through the picture graph above with children. Point out that each colored circle stands for one child, so to find the number of children that chose a certain color, they need to count the colored circles in the row for that color.

Look at the picture graph on student page **IIN43** with children. Read the title of the picture graph and point out that each picture stands for one child's choice. Point out that rose, daisy, and tulip are different kinds of flowers and the different pictures represent the kinds of flowers. Work through the questions with them.

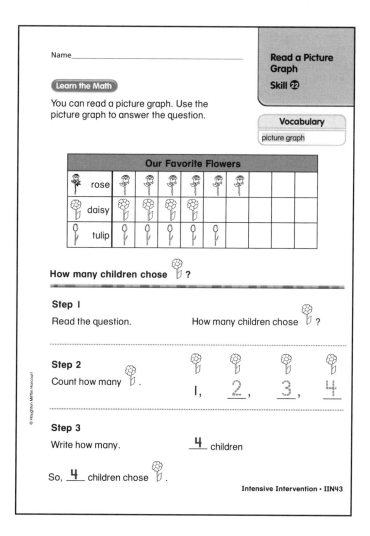

• **What is the question you need to answer?** How many children chose daisy?

• **Which row of the graph shows the information needed to answer the question?** the second row, the row for daisy

• **How many daisies are in the second row?** 4 daisies

• **So, how many children chose daisy as their favorite flower?** 4 children

Have children trace the numbers to count how many chose daisy. Then have them write the number. Make sure children understand how to read a picture graph.

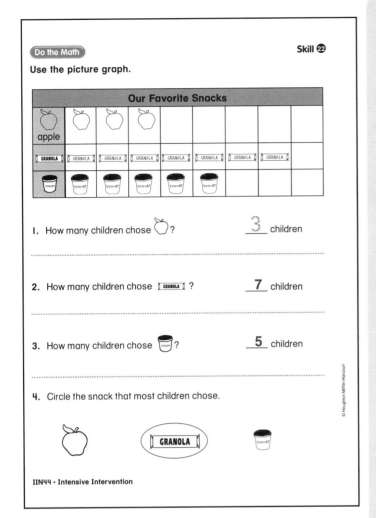

Skill 22

Do the Math

Use the picture graph.

Our Favorite Snacks							
apple	🍎	🍎	🍎				
GRANOLA	GRANOLA	GRANOLA	GRANOLA	GRANOLA	GRANOLA	GRANOLA	
yogurt	yogurt	yogurt	yogurt	yogurt			

1. How many children chose 🍎 ? ___3___ children

2. How many children chose GRANOLA ? ___7___ children

3. How many children chose yogurt ? ___5___ children

4. Circle the snack that most children chose.

🍎   (GRANOLA)   yogurt

IIN44 • Intensive Intervention

© Houghton Mifflin Harcourt

Do the Math

Explain to children that they will use the information in the picture graph on student page IIN44 to answer the questions. Remind them that each picture stands for one child's favorite snack. Have them mark each picture in a row as they count it.

Talk Math

• **What does the first question ask you to find?** how many children chose apple as their favorite snack

• **How will you find how many children chose each snack?** Count the pictures in each row.

Check

Ask: **How can you find which snack the most children chose without counting each picture?** Possible answer: Compare the rows to see which one is the longest.

Alternative Teaching Strategy

Drawing Pictures

Objective To understand how a picture graph can be used to answer questions

Materials drawing paper, crayons

• Ask children to think about which of the following three subjects is their favorite: math, science, or reading.

• Distribute drawing paper and crayons to each child. Ask them to draw a picture to illustrate their choice. For example, they might draw numbers for math, the Earth for science, or a book for reading.

• Sketch the outline of a graph on the board.

Favorite Subject							
math							
science							
reading							

• Call on children to tape their pictures in the appropriate row of the graph. Help children tape the pictures in the appropriate rows, from left to right.

• Once the picture graph is complete, ask questions about the information. Ask: **How many children chose science?** Answers may vary. **How many children chose reading?** Answers may vary. **How many children chose math?** Answers may vary. **How can you tell which subject most children like best?** Look for the row with the most pictures.

• Continue to ask other comparison questions about the picture graph, such as, **Did more children choose science or reading?** Answers may vary.

Intensive Intervention • IIN44

© Houghton Mifflin Harcourt

Make a Prediction
Skill ㉓

Objective
To make a prediction to solve problems

Materials
tiles or counters of two different colors, paper bag

Vocabulary
more likely
prediction

Pre-Assess
Draw a figure on the board that looks like a bag. Inside the bag, draw 4 squares and 2 triangles. Say: **This is a bag. Inside the bag are triangles and squares.** Ask: **How many squares are in the bag?** 4 squares **How many triangles are in the bag?** 2 triangles **If I could reach into this bag without looking and pull out one shape, which shape am I more likely to pick?** a square **Why am I more likely to pick a square than a triangle?** because there are more squares in the bag

Common Misconception
- Children may not understand that you are more likely to pick an object from a bag if there is more of that object than another object in the bag.

- To correct this, place 1 blue tile and 8 white tiles in a bag. On the board, make a tally table to keep track of the blue and white tiles you choose from the bag. Have a child choose a tile and mark its color on the table. Replace the tile after each choice. Discuss with children how many times a blue tile was chosen and how many times a white tile was chosen. Emphasize that if there is more of one kind of object in a bag, it is more likely to be chosen.

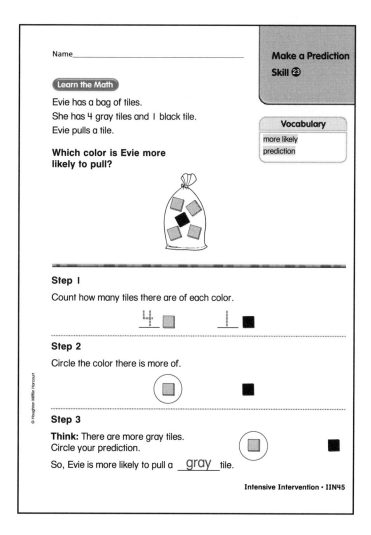

Name_____

Make a Prediction
Skill ㉓

Learn the Math

Evie has a bag of tiles.
She has 4 gray tiles and 1 black tile.
Evie pulls a tile.

Which color is Evie more likely to pull?

Vocabulary
more likely
prediction

Step 1
Count how many tiles there are of each color.

Step 2
Circle the color there is more of.

Step 3
Think: There are more gray tiles.
Circle your prediction.
So, Evie is more likely to pull a ___gray___ tile.

Intensive Intervention • IIN45

Learn the Math

Have children use tiles to model the problem on student page IIN45 as you work through it together.

Talk Math

Have children place 4 gray tiles and 1 black tile in a bag.

- **How many gray tiles are there?** 4 gray tiles
  How many black tiles are there? 1 black tile

- **Are there more gray tiles or more black tiles?** more gray tiles

Have children trace the number of each tile, circle the tile there is more of, and circle the prediction. Make sure they understand that it is more likely for Evie to choose a gray tile than a black tile. Then have them pull one tile from their bag. Have children tell if the tile they pulled from their bag matched their prediction. Point out that a black tile can be pulled from the bag even though there are fewer black tiles than gray tiles.

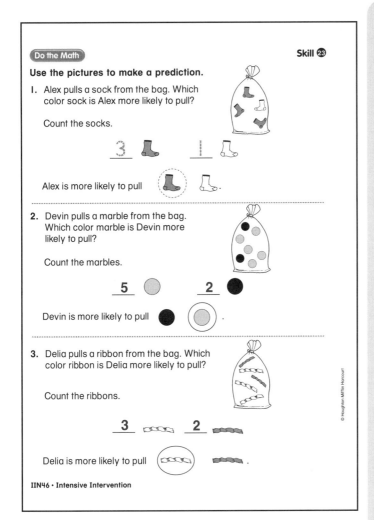

Do the Math

Have children look at student page **IIN46.** Tell them they will make a prediction based on the objects in each bag. Point out that they will begin by counting the objects by color.

Talk Math

- **Why do you need to know how many objects there are of each color?** to find out which color has more objects

- **Which color of object are you more likely to pull from a bag? Explain.** You are more likely to pull the color with the most objects because there are more of them.

Check

Ask: **When you pick an object from a bag without looking, are you more likely to pick the object that there are more of or the object that there are fewer of?** the object that there are more of

Alternative Teaching Strategy

Make a Prediction

Objective To make and test predictions

Materials a paper bag, white and green tiles or red and blue connecting cubes

- Tell children they will make a prediction then act out the problem to test their predictions.

- Have children count with you as you place 5 green tiles and 2 white tiles in a paper bag. On the board, make a two-column tally table. Title the first column *green* and the second column *white.*

- Ask: **How many green tiles are in the bag?** 5 green tiles **How many white tiles are in the bag?** 2 white tiles **Are there more green tiles or white tiles in the bag?** There are more green tiles.

- Say: **Let's make a prediction.** Ask: **If I pull one tile from the bag without looking, which color am I more likely to pull?** a green tile

- Say: **Let's test our prediction.** Have a child pull one tile from the bag. Make a tally mark on the tally table for the color of the tile. Continue having children pull tiles and tally their colors. They should replace the tile after each pull.

- Look at the tally table with children. Ask: **Which color tile was pulled more times: green or white?** Possible answer: green **Was the prediction that you were more likely to pull a green tile correct?** Possible answer: Yes

- You may wish to repeat the activity with different numbers and combinations of tiles.

Objective
To show equal value using pennies, nickels, or dimes

Materials
play pennies, nickels, and dimes

Vocabulary
penny
nickel
dime

Pre-Assess

Show children a nickel. Ask: **What coin is this?** a nickel **What is the value of a nickel?** 5 cents Show children a penny. Ask: **What coin is this?** a penny **What is the value of a penny?** I cent **How many pennies do you need to show the value of a nickel?** 5 pennies Show children a dime. Ask: **What coin is this?** a dime **What is the value of a dime?** I0 cents **How many pennies do you need to show the value of a dime?** I0 pennies Say: **Remember a nickel is worth 5 cents. Ask: How many nickels do you need to show the value of a dime?** 2 nickels

Common Misconception

• Children may incorrectly identify coins and their values.

• To correct this, show children a penny, a nickel, and a dime. Review the characteristics of each coin with children and discuss each coin's value. Provide children with play coins to help them count. Point out that even though a dime is smaller than a nickel it has a greater value.

Learn the Math

Give children I dime, 2 nickels, and I0 pennies. Discuss the value of the coins on student page **IIN47**. Have children use coins to model equal value.

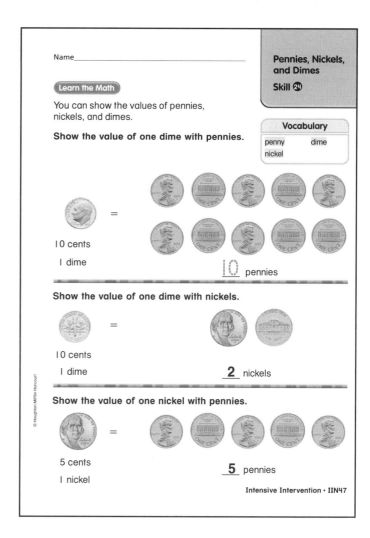

Name_____

Pennies, Nickels, and Dimes
Skill 24

Learn the Math

You can show the values of pennies, nickels, and dimes.

Vocabulary	
penny	dime
nickel	

Show the value of one dime with pennies.

I0 cents
I dime
= ____ 10 pennies

Show the value of one dime with nickels.

I0 cents
I dime
= **2** nickels

Show the value of one nickel with pennies.

5 cents
I nickel
= **5** pennies

Intensive Intervention · IIN47

Talk Math

• **Which coin is a dime?** Hold up a dime and have children do the same.

• **What is the value of a dime?** I0 cents

• **Which coin has the value of I cent?** a penny Hold up a penny and have children do the same.

• **How many pennies do you need to make I0 cents?** I0 pennies

• **So, how many pennies equal the value of I dime?** I0 pennies

Repeat the process by showing equal value for a dime using 2 nickels and a nickel using 5 pennies. Have children trace the number of coins in the first problem and write the number of coins in the remaining problems.

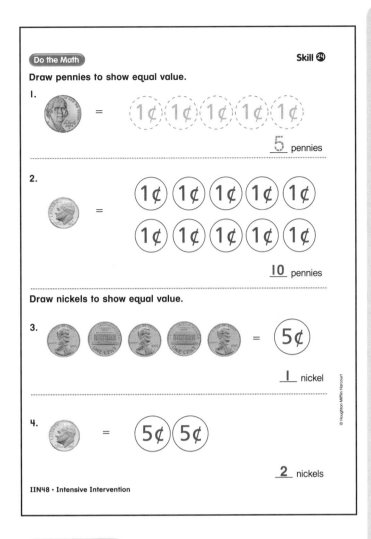

Do the Math

Draw pennies to show equal value.

1. = 1¢ 1¢ 1¢ 1¢ 1¢

   __5__ pennies

2. = 1¢ 1¢ 1¢ 1¢ 1¢ / 1¢ 1¢ 1¢ 1¢ 1¢

   __10__ pennies

Draw nickels to show equal value.

3. = 5¢

   __1__ nickel

4. = 5¢ 5¢

   __2__ nickels

IIN48 · Intensive Intervention

Alternative Teaching Strategy

How Many Ways?

Objective To show equal value using pennies, nickels, or dimes

Materials play money of pennies, nickels, and dimes

• Give each child or pair of children 10 pennies, 2 nickels, and 1 dime.

• Draw a picture of a pencil on the board. Draw a price tag of 10 cents on the pencil. Explain to children that you want to know how many different ways you can pay for the pencil.

• Ask: **How much does the pencil cost?** 10 cents **How many pennies would you need to buy the pencil?** 10 pennies **How many nickels would you need to buy the pencil?** 2 nickels **How many dimes would you need to buy the pencil?** 1 dime **If you wanted to use nickels and pennies to make 10 cents, how many nickels and pennies would you use?** 1 nickel and 5 pennies

• Have children draw all the different ways to make 10 cents.

• Repeat the activity with a five cent price tag. Point out that there are only two ways to make 5 cents, 1 nickel or 5 pennies.

Do the Math

Have children look at student page IIN48. Explain to children that they will draw pennies or nickels to show equal value for the coins. Tell them to read carefully to find out what coins to draw and then think about how many of those coins they need to equal the other coin.

Talk Math

• **What coins will you draw for Problems 1 and 2?** pennies

• **What coins will you draw for Problems 3 and 4?** nickels

Check

Ask: **How do you how many pennies are equal to 1 dime?** Possible answer: I know that a dime is worth 10 cents and I know that a penny is worth 1 cent so a dime is worth 10 pennies.

Objective
To skip-count by fives and tens using pictures

Materials
hundred charts

Pre-Assess

Draw a group of 5 circles on the board. Ask: **How many circles are in the group?** 5 Next to it, draw two more groups of 5 circles. Ask: **How many circles are in each of these groups?** 5 Say: **Skip-count by fives to find how many circles there are in all.** Ask: **How many circles are there in all?** 15 Next, draw a group of 10 stars. Ask: **How many stars are in the group?** 10 Draw two more groups of 10 stars. Say: **Skip-count by tens to find how many stars there are in all.** Ask: **How many stars are there in all?** 30

Common Misconception

- Children may skip a number when skip-counting.

- To correct this, have children use a hundred chart to skip-count. You may wish to have them mark each number as they count it. Tell them to make sure they have counted every 5 or 10 they need to.

Learn the Math

Review counting by fives from 5 to 100 with children using a hundred chart.

Direct children to look at the pictures of the grapes on student page **IIN49**.

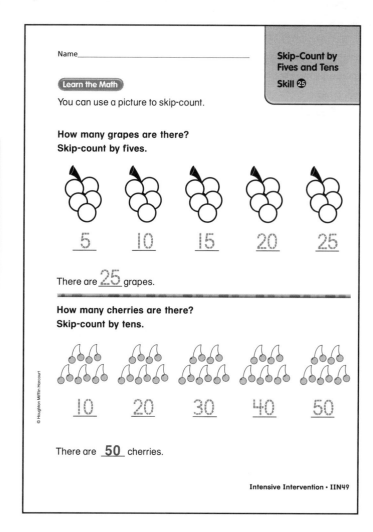

Learn the Math

You can use a picture to skip-count.

**How many grapes are there?
Skip-count by fives.**

5   10   15   20   25

There are 25 grapes.

**How many cherries are there?
Skip-count by tens.**

10   20   30   40   50

There are 50 cherries.

Intensive Intervention • IIN49

Talk Math

- **How many grapes are there in each group?** 5

- **Skip-count by fives. Put your finger on each group of grapes as you count: 5, 10, 15, 20, 25.**

- **How many grapes are there in all?** 25

Have children trace over the numbers of grapes. Then use a hundred chart to review skip-counting by tens from 0 to 100 before repeating the questions for the second problem.

© Houghton Mifflin Harcourt

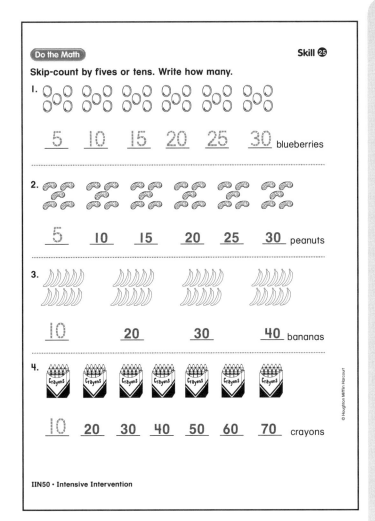

Do the Math   Skill **25**

Skip-count by fives or tens. Write how many.

1. 5   10   15   20   25   30 blueberries

2. 5   10   15   20   25   30 peanuts

3. 10   20   30   40 bananas

4. 10   20   30   40   50   60   70 crayons

IIN50 · Intensive Intervention

Do the Math

Have children look at student page **IIN50**. Explain that they will skip-count by fives or tens to find how many there are in all. If there are 5 in each group, they will skip-count by fives. If there are 10 in each group, they will skip-count by tens.

Talk Math

- **How do you skip-count by fives?** 5, 10, 15, 20, 25, 30,…

- **How do you skip-count by tens?** 10, 20, 30, 40, 50,…

Check

Ask: **When should you skip-count by fives?** When each group has 5 items. **When should you skip-count by tens?** When each group has 10 items.

Alternative Teaching Strategy

By Fives and Tens

Objective To skip-count by fives and tens

Materials counters

- Give each pair of children 40 counters.

- Have children arrange their counters in groups of 5. When they are finished, they should have 8 groups.

- Ask: **How many counters are in each group?** 5 **Will you skip-count by fives or by tens to find how many counters there are in all?** fives

- Say: **When you skip-count by fives, you add 5 to the last number you counted.** Ask: **How many fingers do you have on one hand?** 5

- Say: **Let's skip-count the counters by fives. How many are in the first group?** 5 **So we start with 5. Place your hand between the first and second groups of counters to remind you to add 5.** Ask: **What number do you say as you count the second group?** 10 Say: **Place your hand between the second and third groups. What number do you say as you count the third group?** 15

- Continue this process until children find how many counters there are in all. Then have them rearrange their counters into groups of ten.

- Repeat the activity, but this time have them place both hands between each group of counters to remind them to add 10 instead of 5.

Count Money
Skill ㉖

Objective
To find values of sets of coins including pennies, nickels, and dimes

Materials
play money of pennies, nickels, and dimes

Vocabulary
penny
nickel
dime

Pre-Assess
Give each child five pennies, four nickels, and three dimes. Have them place five pennies in a row on their desks. Ask: **What is the value of each penny?** one cent **What is the value of the pennies?** five cents Have them make a row of four nickels. Ask: **What is the value of each nickel?** five cents **What is the total value of the nickels?** 20 cents Have them make a row of three dimes. Ask: **What is the value of each dime?** 10 cents **What is the total value of the dimes?** 30 cents Next, ask them to place two dimes, three nickels, and one penny in a row. Ask: **What is the total value of the coins?** 36 cents

Common Misconception
- Children may not recognize coins and their values.

- To correct this, make a chart on the board listing the names of each coin and their values. Give each child a penny, a nickel, and a dime. Compare the characteristics of each coin with children so they learn to recognize the differences.

Learn the Math

Give children pennies, nickels, and dimes to model each of the problems on student page IIN51.

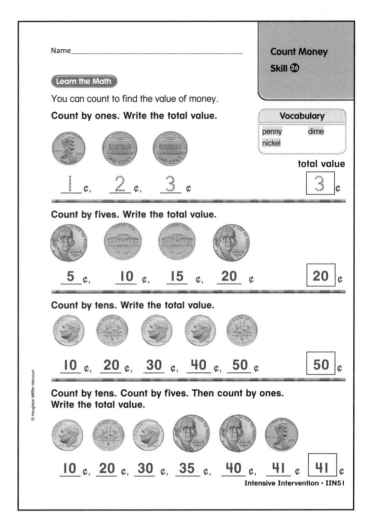

Talk Math

- **What kind of coins are in the first problem?** pennies **What is the value of one penny?** one cent

- Have children point to each penny as you count with them. Say: **Count with me by ones: I cent, 2 cents, 3 cents.** Ask: **What is the total value?** three cents

- **What kind of coins are in the second problem?** nickels **What is the value of one nickel?** five cents

Have children point to each nickel as you count with them. Say: **Count with me by fives: 5 cents, 10 cents, 15 cents, 20 cents.** Ask: **What is the total value?** 20 cents

Have children write the numbers for each amount. Repeat with similar questions for the third and fourth problems.

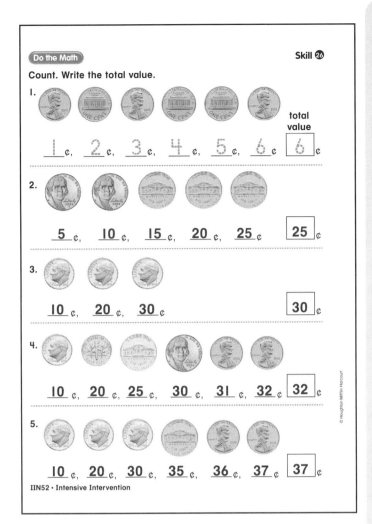

How Much Does It Cost?

Objective To find amounts of money using pennies, nickels, and dimes

Materials play or real pennies, nickels, and dimes; classroom objects such as crayons, pencils, markers, scissors, and erasers

- Set up a pretend store using classroom objects for sale such as crayons, pencils, markers, scissors, and erasers. Number each item.

- Next to each item, instead of a price tag, arrange coins to show the price of each item. For example, next to the pencils place five nickels; next to the scissors place five dimes; next to the eraser place 10 pennies; next to the stapler, place five dimes, one nickel, and one penny. Arrange mixed coins in order: dimes, nickels, and pennies.

- Have children work in pairs. Ask them to visit the store, write down the number of each item and how much it costs.

- Before they begin, review with children how to count coins. Ask: **What do you count by when counting pennies?** ones Ask: **What do you count by when counting nickels?** fives Ask: **What do you count by when counting dimes?** tens

- After children find the cost of each item, have them make pretend purchases with exact change using play or real coins.

Do the Math

Have children look at student page **IIN52**. Point out to children that in Problems 1–3, they will count only pennies, only nickels, or only dimes. In Problems 4 and 5, there is a mix of dimes, nickels, and pennies.

Talk Math

- **What do you count by when counting pennies?** Count by ones.

- **What do you count by when counting nickels?** Count by fives.

- **What do you count by when counting dimes?** Count by tens.

Check

Ask: **When you count a group of pennies, nickels, and dimes, which coin do you start with? Explain.** Possible answer: Start with dimes. It is easier to start with the coins of greatest value.

Objective
To find values of sets of coins including pennies, nickels, dimes, and one quarter

Materials
play money of pennies, nickels, dimes, and quarters

Vocabulary
quarter

Pre-Assess

Give each child one quarter, two dimes, two nickels, and two pennies. Ask them to place the quarter in the middle of their desks. Ask: **What is the value of a quarter?** 25 cents Next, have them add a dime to the right of the quarter. **What is the total value of the coins?** 35 cents Then have them add the second dime. **What is the value of the coins now?** 45 cents Have them add a nickel to the coins. Have children count the coins stating the total value once they add the nickel. **What is the total value of the coins now?** 50 cents Have them add the second nickel. **What is the total value of the coins now?** 55 cents Have them add both pennies to the set of coins. Ask: **What is the value of the money now?** 57 cents

Common Misconception

- Children may automatically assume that the next coin after a quarter is always a dime or that the next coin after a dime is always a nickel and therefore count incorrectly.

- To correct this, point out that the order in which they should count coins is quarter, dime, nickel, penny. Explain that does not mean that each of those coins will always be included. A set of coins may include only a quarter and two nickels, for example. Remind them to look carefully at each coin to be sure they know its value.

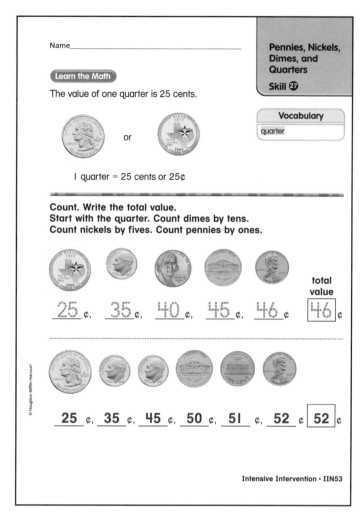

Learn the Math

Give each child one quarter and several nickels, dimes, and pennies to model each of the problems on student page **IIN53**.

Talk Math

- **How many cents is a quarter worth?** 25 cents

- **What is the first number you will say when you begin counting with a quarter?** 25

- **What is a dime worth?** 10 cents

Have children write the numbers for each amount. Continue to ask questions to guide children as they find the value of the sets of coins.

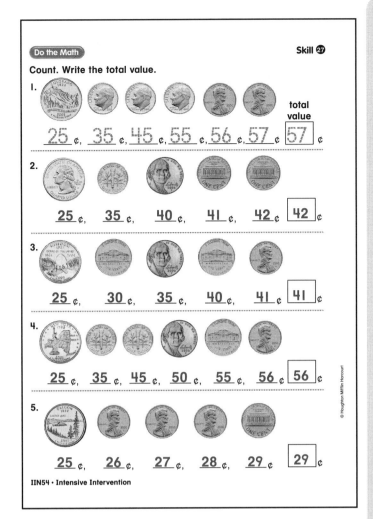

Count. Write the total value.

1. 25 ¢, 35 ¢, 45 ¢, 55 ¢, 56 ¢, 57 ¢   total value [57] ¢

2. 25 ¢, 35 ¢, 40 ¢, 41 ¢, 42 ¢   [42] ¢

3. 25 ¢, 30 ¢, 35 ¢, 40 ¢, 41 ¢   [41] ¢

4. 25 ¢, 35 ¢, 45 ¢, 50 ¢, 55 ¢, 56 ¢   [56] ¢

5. 25 ¢, 26 ¢, 27 ¢, 28 ¢, 29 ¢   [29] ¢

IIN54 • Intensive Intervention

© Houghton Mifflin Harcourt

Alternative Teaching Strategy

How Much Is It?

Objective To find values of sets of coins including pennies, nickels, dimes, and one quarter

Materials play or real pennies, nickels, dimes, and quarters

• Have children work with a partner. Give each pair of children one quarter, three dimes, three nickels, and three pennies.

• Say: **One partner will arrange one quarter and some other coins in a row on your desk. Remember to arrange the coins in order so the quarter is first, followed by dimes, nickels, and then pennies.**

• Before they begin, review with children how to count coins. Ask: **What number do you start with when counting a quarter?** 25 **What do you count by when counting dimes?** tens **What do you count by when counting nickels?** fives **What do you count by when counting pennies?** ones

• Tell partners to each count the money and write the value on a small piece of paper that their partner cannot see. When they are ready, have them turn over their papers at the same time to see if they agree. If they do not agree, have them work together to find the correct value.

• Have partners continue the activity, taking turns arranging coins.

Do the Math

Have children look at student page **IIN54**. Explain to children that they will find the value of sets of coins including quarters, dimes, nickels, and pennies. Point out that not every problem has every kind of coin. Remind them to look at each coin carefully.

Talk Math

• **What number do you start with when you count a quarter?** 25

• **Which coins do you count by tens?** dimes

• **Which coins do you count by fives?** nickels

• **Which coins do you count by ones?** pennies

Check

Ask: **What amount of money do you have if you have one quarter, one dime, one nickel, and one penny?** 41 cents

© Houghton Mifflin Harcourt

Use a Clock
Skill 28

Objective
To understand how to read a clock to the nearest hour

Materials
analog demonstration clock

Vocabulary
clock
minute hand
hour hand

Pre-Assess

Draw an analog clock on the board or use a demonstration clock. Show 8:00. Ask: **What number does the minute hand point to?** 12 **What number does the hour hand point to?** 8 **What time does the clock show?** 8 o'clock Show 2:00. Ask: **What number does the minute hand point to?** 12 **What number does the hour hand point to?** 2 **What time does the clock show?** 2 o' clock Continue showing children other times and asking children to tell the times.

Common Misconception

• Children may confuse the hour hand and the minute hand.

• To correct this, show 12:00 on a demonstration clock. Point out that the hour hand is shorter than the minute hand. Explain that the minute hand touches the numbers on a clock, but the hour hand does not.

Learn the Math

You may wish to use a demonstration clock to show 7:00 as you guide children to use a clock on student page **IIN55**.

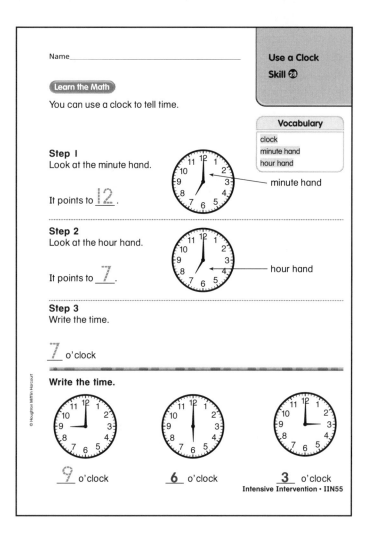

Talk Math

• **What are the numbers on the clock?** 1, 2, 3, 4, 5, 6, 7, 8, 9, 10, 11, and 12

• **Put your finger on the minute hand. What number is it pointing to?** 12

• **Put your finger on the hour hand. What number is it pointing to?** 7

• **What is the time?** 7 o'clock

Repeat the questions to guide children in telling the time on the clocks at the bottom of the page.

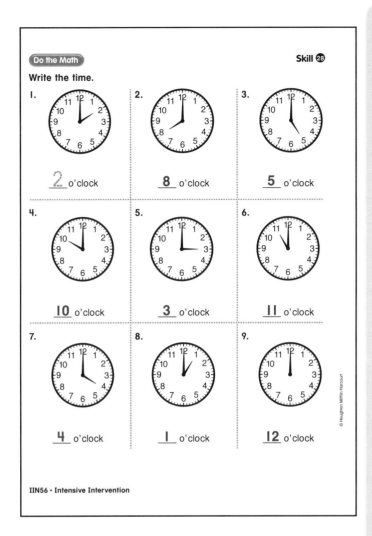

Write the time.

Skill 28

1. **2** o'clock

2. **8** o'clock

3. **5** o'clock

4. **10** o'clock

5. **3** o'clock

6. **11** o'clock

7. **4** o'clock

8. **1** o'clock

9. **12** o'clock

IIN56 · Intensive Intervention

Do the Math

Have children look at student page **IIN56**. Tell them that they will write the time to the hour shown on each clock. Point out that the minute hand on each clock points to 12. Children will tell the time on each clock by finding the number to which the hour hand points.

Talk Math

- **Which hand do you look at to find the time to the hour?** the hour hand

- **To which number does the minute hand point when the time is exactly to the hour?** 12

Check

Ask: **When it is 4 o'clock, to which number does the hour hand point?** 4 **To which number does the minute hand point?** 12

Alternative Teaching Strategy

What Time Is It?

Objective To use a clock to tell time to the hour

Materials analog demonstration clock, clocks for each child

- Have children work in pairs. Give each child a demonstration clock.

- Explain to children that they will use the clocks to practice telling and showing time to the hour. Show 6:00 on a demonstration clock. Ask: **To which number is the minute hand pointing to?** 12 Remind children that if the time is to the hour, the minute hand will always point to 12.

- Ask: **To which number is the hour hand pointing?** 6 **What time is it?** 6 o'clock

- Say: **Now, work with your partner. Show a time to the hour on your clock. Remember to point the minute hand to 12 and the hour hand to one of the numbers on the clock. Then ask your partner to tell what time it is.** Give children a minute to do so.

- Say: **Now switch. Whoever showed the time will now tell the time that their partner shows.**

- Continue the activity until most pairs have shown many of the times to the hour.

Objective
To estimate whether an activity would take about a minute or about an hour, or several minutes or several hours

Vocabulary
minute
hour
estimate

Pre-Assess
Write the following on the board:

about 1 minute

about 1 hour

Say: **I am going to name an activity. Tell me if it would take about one minute or about one hour to complete the activity.** Say: **Eat at a restaurant.** about 1 hour **Put on a coat.** about 1 minute **Pour a glass of water.** about 1 minute **Bake a cake.** about 1 hour

Common Misconception

• Children may confuse one minute and one hour.

• To correct this, point out the second hand on a clock. Have children sit silently for exactly one minute. Then explain that one hour is 60 minutes. So the second hand would have to go around the clock 60 times to equal one hour. Discuss with children things they think they could do in one minute and things they think they could do in one hour.

Learn the Math

You may wish to give children points of reference for how long activities take throughout the day. For example, you may tell them that it takes about 20 minutes to eat lunch.

Guide children through the problem at the top of student page **IIN57**. Demonstrate that it takes a short amount of time to tie a shoe. Then discuss

Name_____

Minutes and Hours Skill 29

Learn the Math
You can estimate about how long it takes to do something.

Vocabulary

minute   estimate
hour

tie your shoes

about 1 minute

cook dinner

about 1 hour

About how long does it take? Circle your answer.

exercise in gym class

about one minute

(about one hour)

brush your teeth

(about two minutes)

about two hours

Intensive Intervention • IIN57

how it takes a longer time, about an hour, to cook dinner. Point out that these are estimates of the time it takes to do activities. Then direct children to look at the pictures at the bottom of the page.

Talk Math

• **What are the children in the first picture doing?** Exercising in gym class.

• **Think about how long one minute is. Do you think gym class lasts about one minute?** No, gym class is longer than one minute.

• **Think about how long one hour is. Do you think gym class lasts about one hour?** yes

Repeat the questions to guide children through the second problem. Point out that the answer choices are not one minute and one hour, but two minutes and two hours.

Do the Math

Do the Math

Have children look at student page **IIN58**. Explain to children that they will choose between two amounts of time to tell about how long each activity takes. Tell them to think about each choice carefully.

Talk Math

• **Will an activity that takes about a minute be writing your name or writing a story?** writing your name

• **Does an activity that takes about an hour take a long time or a short time?** a long time

Check

Ask: **What is an activity that takes about one minute?** Possible answer: putting on my socks
What is an activity that takes about one hour? Possible answer: playing a game

Alternative Teaching Strategy

Act It Out

Objective To act out activities to determine whether they will take about one minute or about one hour

• Explain to children that they will estimate whether the following activities take about one minute or about one hour. Then they will act out the activities to see if their estimate is correct.

• Write the following activites on the board:

> Make a bowl of cereal.
> Clean your room.
> Pour a glass of water.
> Do an art project.

• Ask: **About how long do you think it would take to make a bowl of cereal?** about one minute Say: **Okay, when I say "start," pretend you are putting together the ingredients to make a bowl of cereal. Don't forget the milk. When you are finished, put your hands on your lap. Do not rush.** Time children to see about how long it would take to make a bowl of cereal.

• Discuss the outcome with children. Lead them to realize that making a bowl of cereal takes about one minute.

• Repeat with other activities such as erasing the board or writing the alphabet. For the activities that would take about an hour, have children begin pretending and stop them after one minute. Ask: **How much did you get done in one minute?** very little Ask: **Do you think your estimate of one hour is correct?** yes

Sort by Color, Size, and Shape
Skill ③⓪

Objective
To sort objects according to color, size, and shape

Pre-Assess

Draw two triangles and a square on the board. Ask: **Which shape does not belong?** the square Cross out the square. Draw 3 small circles and 1 large circle. Ask: **Which shape does not belong?** the large circle Cross out the large circle. Point to the group of circles. Ask: **How are these sorted – by color, size, or shape?** size Draw 3 squares and shade 2 of them. Ask: **Which shape does not belong?** the square that is not shaded

Common Misconception

• Children may have difficulty deciding which object does not belong in a group if the group shares more than one attribute.

• To correct this, draw 4 rectangles on the board. Shade 3 of them. Point out that if you sort them by color, the rectangle which is not shaded does not belong, but if you sort them by shape, all of the shapes belong in the group. Tell children to look for the object which has one attribute that is different from all of the other objects. Emphasize that it may still share some other attributes, but one is different.

Learn the Math

Work through the different ways to sort objects on student page **IIN59** with children.

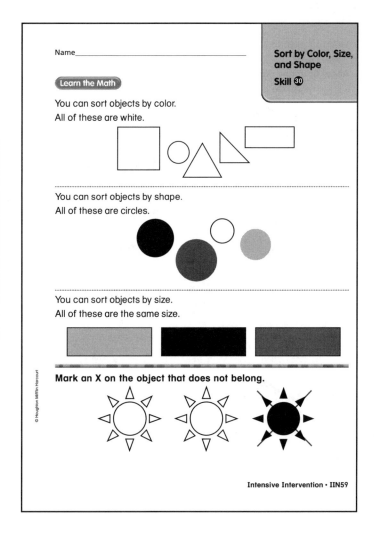

Talk Math

• **Direct children to the group of white shapes.** Ask: **Are all of these the same shape?** no

• **Are they all the same size?** no

• **Are they all the same color?** yes

• **How are they sorted?** by color

Repeat with the objects sorted by shape and size, changing the questions accordingly. Then ask the same series of questions to help children solve the problem at the bottom of the page.

Do the Math

Skill 30

Mark an X on the one that does not belong.

1.

2.

Circle the group in which the object belongs.

3.

4.

IIN60 · Intensive Intervention

Do the Math

Have children look at student page **IIN60**. Explain to children that for Problems 1 and 2 they will need to decide which object does not belong to the group. In Problems 3 and 4, they will need to decide in which group the object on the left belongs.

Talk Math

• **Look at Problem 1. Which one does not belong?** the square **How are the three that belong alike?** They are the same shape.

• **Look at Problem 3. How are each of the two groups sorted?** They are sorted by color.

Check

Draw 4 rectangles of different sizes and colors on the board. Ask: **Are these sorted by color, shape, or size?** shape **Draw another object that belongs in this group.** Children should draw a rectangle.

Alternative Teaching Strategy

Sort It Out

Objective To sort objects by color, shape, and size

Materials 4 triangles, 4 rectangles, and 4 circles

In each set, there should be 1 small red shape, 1 small blue shape, 1 large red shape, and 1 large blue shape. The shapes should be large enough so many children can see them.

• Have children sit in a circle on the floor or gather around a table. Explain that they will be sorting by color, shape, and size.

• Ask a volunteer to choose one shape from the group. Discuss the attributes of the shape. Ask: **What shape is it? What color is it? What size is it, small or big?** Answers may vary.

• Say: **Let's sort the shapes.** Ask volunteers to find all of the other shapes that are the same color. Ask: **How did you sort, by shape, color, or size?** by color Have children return the shapes.

• Say: **Let's sort another way.** Ask volunteers to find all of the same shape. Ask: **How did you sort, by color, shape, or size?** by shape

• Say: **Now, let's sort another way.** Ask volunteers to find all of the same size. Ask: **How did you sort, by color, shape, or size?** by size

• Make some random groups where one shape does not belong in the group. Work with children to identify the one that does not belong and have them explain why it does not belong.

Intensive Intervention · IIN60

Objective
To identify three-dimensional shapes

Materials
three-dimensional shapes and everyday objects in the shape of cubes, rectangular prisms, cones, cylinders, spheres, and pyramids

Vocabulary
cone
cube
cylinder
pyramid
rectangular prism
sphere
three-dimensional

Pre-Assess
Gather classroom objects that are shaped like a sphere, a cone, a cylinder, a pyramid, a rectangular prism, and a cube. Place the objects on a table in the front of the classroom. Hold up each object, one at a time, and ask students to identify each three-dimensional shape.

Common Misconception
- Children may not be able to distinguish cubes from other rectangular prisms.

- To help children, display several rectangular prisms including some cubes. Point out that a cube is a special type of rectangular prism— each flat surface is a square and is the same size. To verify, trace each of the flat surfaces of a cube on a sheet of paper, cut them out, and match the sides. Repeat with a rectangular prism that is not a cube.

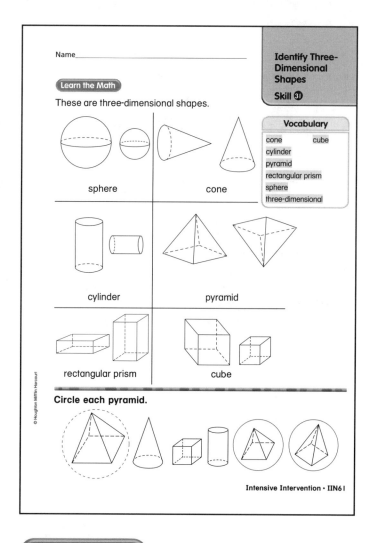

Learn the Math
Display models of each shape and discuss the attributes of each.

Talk Math
- **Look at the sphere. Name an object that is shaped like a sphere.** Possible answer: a ball

- **Look at the cone. Name an object that is shaped like a cone.** Possible answer: an ice cream cone

- **Look at the cylinder. It is flat on the top and bottom, but it will roll on its side. Name an object that is shaped like a cylinder.** Possible answer: a can

Repeat for the pyramid, rectangular prism, and cube.

Guide children to look at the problem at the bottom of student page **IIN61**. Have children identify each shape and circle all of the pyramids.

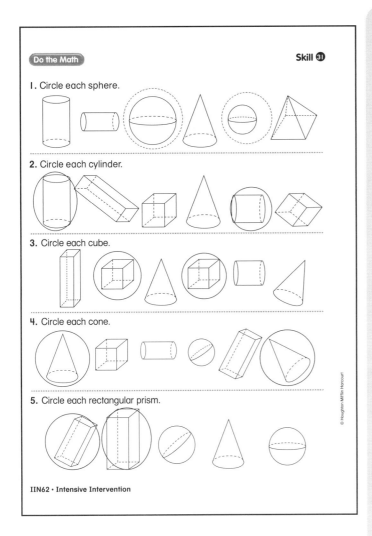

1. Circle each sphere.

2. Circle each cylinder.

3. Circle each cube.

4. Circle each cone.

5. Circle each rectangular prism.

IIN62 · Intensive Intervention

© Houghton Mifflin Harcourt

Do the Math

Have children look at student page **IIN62**. Explain that they will circle all of the shapes of a certain kind.

Talk Math

- **How can you tell a cone from a cylinder?**
  Possible answer: A cone has a point and one flat surface shaped like a circle. A cylinder has flat surfaces shaped like circles.

- **How can you tell a sphere from a cube?** A sphere is round and shaped like a ball. A cube is shaped like a box.

Check

Ask: **What is a cereal box shaped like?** a rectangular prism

Alternative Teaching Strategy

What Is It Shaped Like?

Objective To identify three-dimensional shapes using real-world objects

- Have children work in pairs. Write a table on the board like the one below.

Shape	Real Object
sphere	
cone	
cylinder	
cube	
rectangular prism	

- Say: **Work with your partner to find a real-world object that is like each shape.**

- Give children time to look for real-world objects. Assist them as needed.

- Say: **Name an object that is shaped like a sphere.** Possible answers: globe, soccer ball **Name an object that is shaped like a cone.** Possible answers: traffic cone, ice cream cone **Name an object that is shaped like a cylinder.** Possible answers: pencil holder, cup **Name an object that is shaped like a cube.** Possible answers: tissue box, number cube **Name an object that is shaped like a rectangular prism.** Possible answers: books, shoe box

- As children name real-world objects, write them on the table. Encourage them to see that there are many examples of three-dimensional shapes in the real-world.

Sides and Vertices
Skill ㉜

Objective
To identify and count sides and vertices of two-dimensional shapes

Vocabulary
side

vertex/vertices

Pre-Assess
Draw a five-sided shape on the board. Ask: **How many sides does the shape have?** 5 sides Ask: **How many vertices does the shape have?** 5 vertices Repeat with a triangle and a square.

Common Misconception
• Children may miscount the number of sides and vertices.

• To correct this, have children draw a circle around each vertex as they count the number of vertices and trace each side as they count the number of sides. Point out that as they begin to recognize shapes, they will be able to count the number of sides and vertices correctly.

Learn the Math
Have children point to each side and vertex of the square and triangle on student page **IIN63** as you count with them.

Talk Math
• **Look at the first shape.** Children may recognize the shape as a square. **The arrow points to one side of the square. Count all of the sides. How many sides does the square have?** 4 sides

• **A vertex of the square is circled. Count all of the vertices. How many vertices does the square have?** 4 vertices

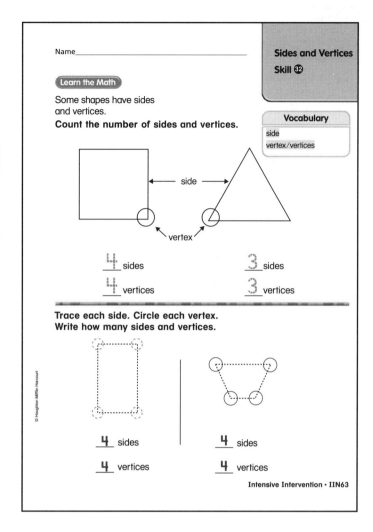

• **Look at the second shape.** Children may recognize the shape as a triangle. **The arrow points to one side of the triangle. Count all of the sides. How many sides does the triangle have?** 3 sides

• **A vertex of the triangle is circled. Count all of the vertices. How many vertices does the triangle have?** 3 vertices

Guide children through the problems at the bottom of the page. Have children trace the sides of the shapes and circle the vertices. Then have them write the number of sides and vertices.

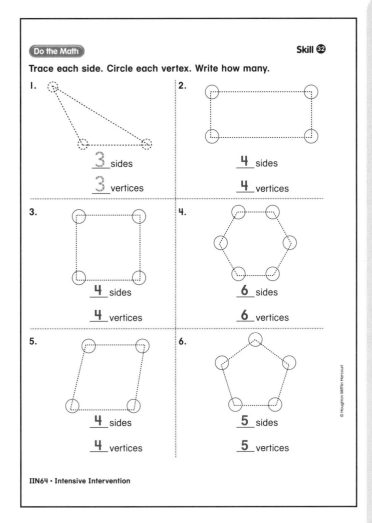

Do the Math

Have children look at student page **IIN64**. Tell children that they will trace each side, circle each vertex, then count and write the number of sides and vertices.

Talk Math

• **What does a side look like?** It is a straight line.

• **What does a vertex look like?** It is where two sides meet.

• **How can you find the number of sides and vertices of a shape?** Look at the shape and count the number of sides and vertices.

Check

Say: **Draw a shape with 4 sides.** Check children's drawings. **How many vertices does it have?** 4 vertices

Alternative Teaching Strategy

Side and Vertex Hunt

Objective To identify and count sides and vertices of two-dimensional shapes

Materials drawing paper, magazines

• Review with children what sides and vertices of a shape are. Then tell them that they will look for shapes in the classroom and count their sides and vertices.

• Point out that they can sometimes find shapes on the sides of objects.

• Say: **Look around the classroom. Find 3 shapes with 4 sides and 4 vertices. Write down or draw each shape that you find.** Point out that the sides of the shape do not have to be equal.

• Give children time to find and write down shapes. Then have volunteers describe the shapes they found.

• Repeat the activity for shapes with 3 sides and 3 vertices. If children have difficulty finding shapes, have them look in magazines or draw shapes themselves.

Sort Two-Dimensional Shapes
Skill 33

Objective
To sort and identify shapes based on the number of sides and vertices

Pre-Assess

Draw a square, a triangle, a rectangle, and a circle on the board. Ask: **Which shapes have 4 sides and 4 vertices?** Children should identify the square and the rectangle. **What are the names of those shapes?** square and rectangle Ask: **Which shape has 3 sides and 3 vertices?** Children should identify the triangle. **What is the name of that shape?** triangle

Common Misconception

• Children may confuse a square and a rectangle because they have the same number of sides and vertices.

• To correct this, tell children that a square is a special kind of rectangle. Remind them that a square and a rectangle have the same number of sides and vertices, but the sides of a square are all the same length. Only the opposite sides of a rectangle are the same length.

Learn the Math

Have children point to each side and vertex of the rectangle, square, and triangle on student page **IIN65** as you count the number of sides and vertices with them.

Talk Math

• **How many sides does a rectangle have?** 4 sides
How many vertices does a rectangle have? 4 vertices

• **How many sides does a square have?** 4 sides
How many vertices does a square have? 4 vertices

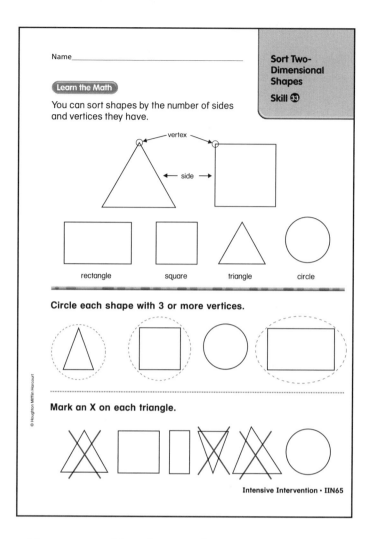

• **How many sides does a triangle have?** 3 sides
How many vertices does a triangle have? 3 vertices

Guide children through the the first problem. Tell them to count the number of vertices on each shape. Tell them to trace the circles around the shapes having 3 or more vertices.

Direct children to refer to the triangle at the top of the page to help them with the second problem. Guide them to see that all triangles have 3 sides and 3 vertices. Have them mark an X on each triangle.

© Houghton Mifflin Harcourt

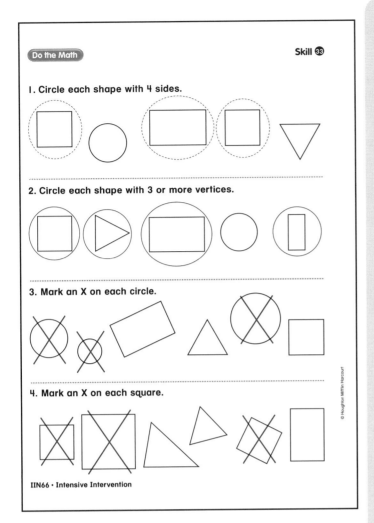

Do the Math

Guide children as they complete Problems 1 – 4 on student page **IIN66**. Read the directions aloud for each problem.

Talk Math

- **Look at Problem 1. What is one way you can sort shapes?** Possible answer: by the number of sides

- **Look at Problem 2. What is another way you can sort shapes?** Possible answer: by the number of vertices

Check

Ask: **What is the name of a shape with 4 sides?** square or rectangle **What is the name of a shape with 3 vertices?** triangle

Alternative Teaching Strategy

Sort It Out

Objective To sort two-dimensional shapes based on the number of sides and vertices

Materials paper, scissors, two-dimensional shapes (see Teacher Resource Book), including circles, squares, and rectangles

- Prepare an assortment of two-dimensional shapes for children to cut out.

- Collect all of the shapes. Place them on the floor or on a table in the middle of the room. Randomly pull one shape from the group.

- Ask such questions as: **How many sides does this shape have? How many vertices does it have? What is the name of this shape?** Answers may vary.

- Ask for a volunteer to find another shape in the group that could be sorted with the first shape. For example, if the first shape is a triangle, have them find another triangle.

- Continue until all of the shapes have been sorted. Give each group of shapes a name based on the number of sides and vertices such as, "Shapes with 4 sides and 4 vertices."

Identify and Copy Patterns
Skill 34

Objective
To identify and copy patterns

Materials
pattern blocks

Vocabulary
pattern

Pre-Assess

Display the following pattern with pattern blocks: hexagon, square, hexagon, square, hexagon, square. Say: **Use pattern blocks to copy the pattern.** Encourage children to describe the pattern either by color or shape.

Common Misconception

• Children may not be able to identify the part of the pattern that repeats.

• To correct this, circle the first part of the pattern (pattern unit) for children. Show them how the pattern repeats. You may wish to introduce the term "pattern unit." Underline each pattern unit so children can see the pattern more clearly.

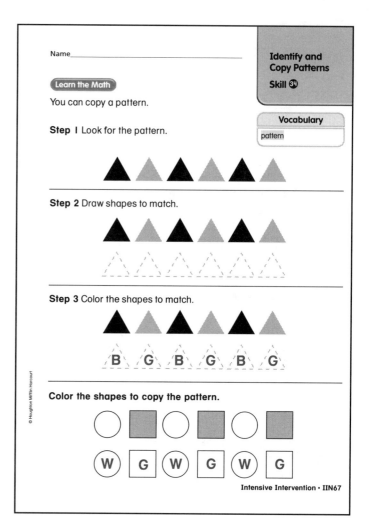

Learn the Math

Help children through the process of identifying and copying a pattern on student page **IIN67**. Help children identify the repeating part (pattern unit) of each pattern.

Talk Math

• **Look at the shapes in Step 1. What do you notice about the pattern?** Possible answers: There are black and gray triangles. There is one black, then one gray, then one black and so on.

• **What part of the pattern repeats?** black triangle, gray triangle

Direct children to Step 2. Have them copy the pattern by tracing the dashed lines for each triangle. Then direct them to Step 3.

• **What color will you make the first triangle to copy the pattern?** black

• **What color will you make the second triangle?** gray

Continue asking children to tell what color each of the remaining shapes in the pattern will be. Then help them with the problem at the bottom of the page to identify the pattern and then color the shapes to copy the pattern.

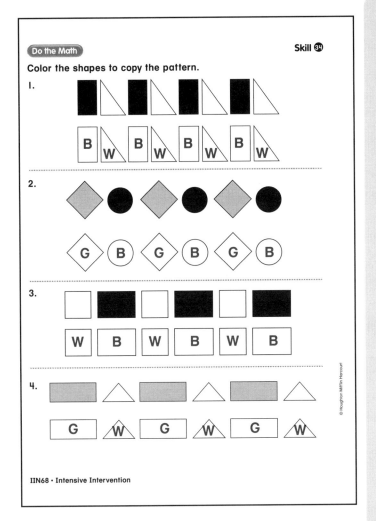

Do the Math

Skill **34**

Color the shapes to copy the pattern.

1.

2.

3.

4.

IIN68 · Intensive Intervention

Do the Math

Have children look at student page **IIN68**. Help them identify the pattern in each row, then color the shapes to copy the pattern.

Talk Math

- **How will you know what color to make the first shape in each pattern you copy?** Look at the first shape of the original pattern and copy the color.

- **How will you know what colors to make the other shapes in the patterns you copy?** Possible answers: Look at each shape and copy the color; identify the part of the pattern that repeats and keep repeating it.

Check

Show the following pattern with pattern blocks: square, triangle, square, triangle, square, triangle. Ask: **What part of this pattern repeats?** square and triangle

Alternative Teaching Strategy

Pattern

Objective To identify and copy patterns

Materials 12 sheets of construction paper (6 blue, 6 yellow)

- Have six children stand in front of the class. Have three of them each hold a yellow sheet of construction paper and three of them each hold a blue sheet of construction paper. Arrange children in a pattern of blue, yellow, blue, yellow, blue, yellow, facing the class.

- Discuss the pattern with children. Ask: **What is the color of the first sheet of construction paper in the pattern?** blue **What is the color of the second sheet of construction paper in the pattern?** yellow Help children to describe the pattern: blue, yellow, blue, yellow, blue, yellow.

- Have six more children stand in a row next to the other children. Say: **Let's copy the pattern.** Ask: **What color should the first child hold?** blue Give the first child a blue sheet of construction paper. Ask: **What color should the second child hold?** yellow Give the second child a yellow sheet of construction paper. Continue until the pattern has been copied.

- Repeat the activity using different colors and different volunteers.

Extend Patterns
Skill ③⑤

Objective
To identify and extend patterns

Materials
two-dimensional shapes

Vocabulary
pattern unit

Pre-Assess

Draw the following six shapes on the board in a row: square, rectangle, square, rectangle, square, rectangle. Say: **Look for a pattern.** Ask: **What is the pattern unit?** square, rectangle **What comes next?** square Continue with other AB patterns.

Common Misconception

• Children may reverse the order of the shapes in the pattern unit.

• To correct this, have children identify the shapes used in the pattern. Then have them say the names of the shapes in the order of the pattern unit.

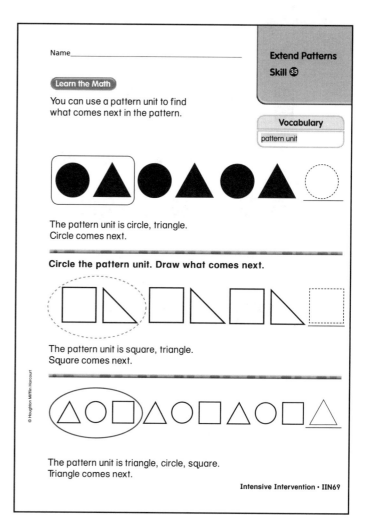

Learn the Math

You can use a pattern unit to find what comes next in the pattern.

Vocabulary

pattern unit

The pattern unit is circle, triangle.
Circle comes next.

Circle the pattern unit. Draw what comes next.

The pattern unit is square, triangle.
Square comes next.

The pattern unit is triangle, circle, square.
Triangle comes next.

Intensive Intervention • IIN69

Learn the Math

Guide children through finding a pattern unit and extending a pattern on student page **IIN69**. Point out that the pattern unit is the shapes in the pattern that repeat over and over. You may wish to have children model each pattern.

Talk Math

• **Look at the pattern. What shapes repeat?** circle, triangle

• **Which shape is first in the pattern?** circle Explain that this shape is first in the pattern unit.

• **So, what is the pattern unit?** circle, triangle

• **Look at the pattern unit. Does a circle or a triangle come next in the pattern?** circle

Have children trace the circle that comes next in the pattern. Repeat the questions for the second problem.

Then, direct children to the third problem.

• **How is this pattern unit different from the one in the first problem?** It has three shapes.

• **What is the pattern unit?** triangle, circle, square

• **What comes next in the pattern?** triangle

Direct children to circle the pattern unit and then draw the triangle to extend the pattern.

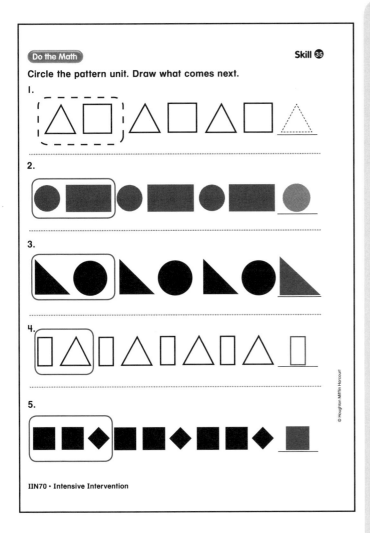

Do the Math Skill 35

Circle the pattern unit. Draw what comes next.

1.

2.

3.

4.

5.

IIN70 · Intensive Intervention

Do the Math

Have children look at student page **IIN70**. Explain to them that they will look at a pattern, circle the pattern unit, and then use the pattern unit to extend the pattern.

Talk Math

• **How do you find the pattern unit?** Look for the group of shapes that repeats.

• **How does the pattern unit help you find what comes next?** When you know the pattern unit, you know the order of the shapes so you can find which comes next.

Check

Ask: **How do you know which shape comes next in a pattern?** Look at the pattern unit to find out which group of shapes repeats, then decide which shape comes next.

Alternative Teaching Strategy

Make a Pattern

Objective To create a pattern

Materials pattern blocks

• Distribute pattern blocks to pairs of children.

• Say: **I will show you how to make a pattern using these shapes. First, I need to create a pattern unit. A pattern unit is a group of shapes that repeats. I will use a square and a triangle for my pattern unit.**

• Display a square and a triangle. Make sure that the colors also repeat in the pattern. Say: **I will keep repeating this pattern unit.** Display three more sets of the pattern unit. Ask: **Now I have made a pattern. What shape comes next if I want to continue the pattern?** square

• Say: **Now one partner will make a pattern and the other will extend it.** Guide children through the process of creating a pattern unit and repeating it. Remind children that the colors also have to repeat in the pattern. After they have created the pattern, say: **Now I want the other partner to show what shape will come next.**

• Have partners switch roles. Repeat this activity several more times to give children time to create and extend patterns.

Compare Lengths
Skill ㊱

Objective
To compare lengths of objects and order them according to size

Vocabulary
longest
shortest

Pre-Assess
Draw three horizontal lines of different lengths on the board. Do not draw them in order of length.

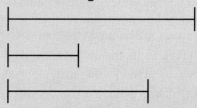

Ask: **Which line is the longest?** the top line **Which line is the shortest?** the middle line Have students order the lines from shortest to longest by numbering the shortest line *1*, the next shortest line *2*, and the longest line *3*.

Common Misconception

- Children may have difficulty understanding how to order objects from shortest to longest.

- To correct this, draw three objects in order from shortest to longest starting from the top. Point out the first and write "shortest" next to it. Point out the last and write "longest" next to it. Then, to show children the order, write "1" next to the shortest object, "2" next to the middle object, and "3" next to the longest object.

Learn the Math

Point out the marks at the beginning of each object. Explain to children that to compare lengths, the beginning of the objects need to be lined up. Have children point to each object on student page **IIN71** as you guide them through comparing the lengths of objects.

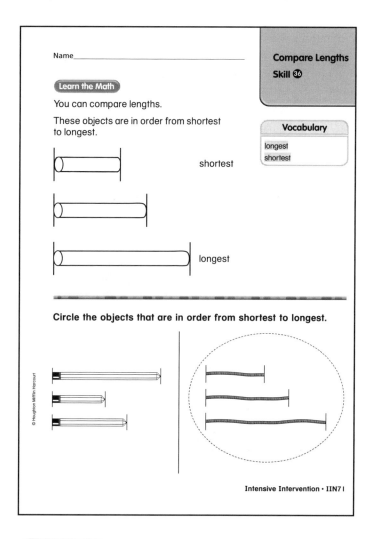

Talk Math

- **Look at the three pieces of chalk. Point to the shortest piece. Why is it shortest?** because it is shorter than the other two from end to end

- **Point to the next piece of chalk. It is not the longest or the shortest. Why?** It is longer than the shortest piece and shorter than the longest piece.

- **Point to the longest piece of chalk. Why is it the longest?** because it is longer than the other two from end to end

At the bottom of the page, have children identify the shortest and longest object in each group. You may wish to have them put an X on the shortest object and underline the longest object. Then have them compare to find which group is ordered correctly.

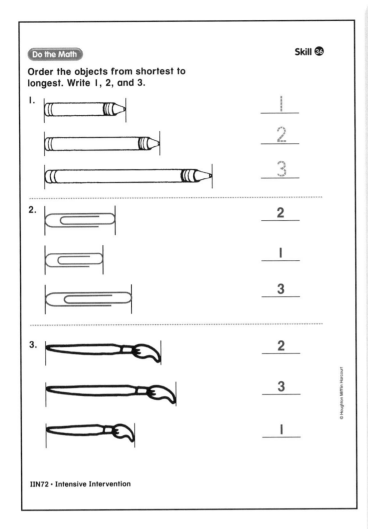

Do the Math

Read aloud the directions on student page **IIN72**.

Talk Math

- **Which crayon is the shortest?** The one on top.

- **Which crayon is the longest?** The one on the bottom.

- **What if you ordered the crayons from longest to shortest? How would the numbers change?** The longest and shortest numbers would be reversed.

Check

Ask: **What happens if you do not line up the beginning of all objects when you compare lengths?** You may not be able to compare them correctly.

Alternative Teaching Strategy
Shorter and Longer

Objective To compare lengths of objects and order them according to size

Materials classroom objects, paper, pencils

- Explain to children that they will compare the lengths of ordinary objects in their classroom.

- Say: **Choose one object that is sitting on your desk or table. Trace your object on a piece of paper.** Give children time to trace their objects.

- Say: **Now I want you to find two other objects in the room. I want you to find one object that is longer than your object, and one object that is shorter than your object. Trace each of those objects on your paper so that the left ends of the objects align with the left end of your first object.** Give children time to find two other objects and trace them. Encourage children to choose flat objects that are easy to trace.

- Say: **Now, compare the lengths of your objects. Circle the one that is the shortest. Draw a line under the one that is the longest.**

- Say: **Now number your objects in order from shortest to longest. Number the shortest object *1*, the next object *2*, and the longest object *3*.**

- Repeat with other objects as time allows.

Temperature
Skill ㉞

Objective
To read temperature on a Fahrenheit thermometer

Materials
Fahrenheit thermometer

Vocabulary
temperature
thermometer

Pre-Assess

Draw a Fahrenheit thermometer from 0°F to 100°F on the board. Fill in a temperature of 60°F. Draw a second Fahrenheit thermometer next to the first. Fill in a temperature of 30°F. Point to the first thermometer. Ask: **What temperature does the thermometer show?** 60°F Point to the second thermometer. Ask: **What temperature does the thermometer show?** 30°F **Which thermometer shows the hotter temperature?** the first thermometer **Which thermometer shows the colder temperature?** the second thermometer

Common Misconception

• Children may not line up the number and the shaded line correctly when they read a thermometer.

• To correct this, review with children how to read a thermometer by placing the edge of a piece of paper on the line so they can read the scale for the temperature.

Learn the Math

Display a Fahrenheit thermometer. Tell children there are different kinds of thermometers but this one is a Fahrenheit thermometer. Say: **A thermometer tells us how hot or cold something is.** Help children read the thermometer by finding the line and the number where the measurement ends. Then explain that the higher the measurement, the hotter the temperature.

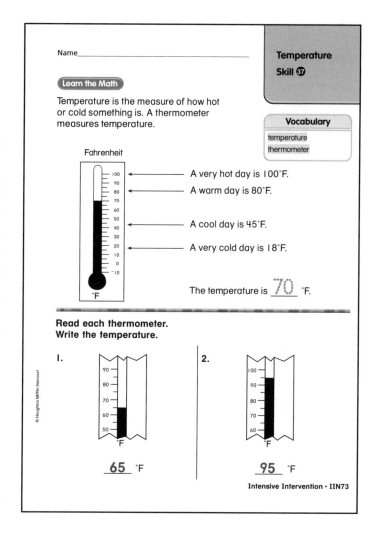

Invite children to share activities they might do on a very hot day or a very cold day.

Direct children to look at the thermometer at the top of student page **IIN73**. Remind children to look at where the measurement line ends and line it up with the number that shows the temperature.

Talk Math

• **What temperature does the thermometer show?** 70°F

• Direct children to the next two thermometers. **What temperature does the left thermometer show?** 65°F

• **What temperature does the right thermometer show?** 95°

Do the Math

Skill ③⑦

Read each thermometer.
Write the temperature.

1. **90** °F

2. **20** °F

3. **55** °F

4. **40** °F

5. **75** °F

6. **15** °F

7. **10** °F

8. **70** °F

9. **35** °F

IIN74 · Intensive Intervention

Do the Math

Have children look at student page **IIN74**. Explain that they need to read the temperature on each thermometer and write the temperature under it. Encourage them to use the edge of a piece of paper to line up the marker with the number.

Talk Math

• **How do you read the temperature on a thermometer?** Find the number where the measurement ends. This is the temperature.

Check

Say: **In the morning, Julia needed a jacket. In the afternoon, she put on shorts.** Ask: **When was the temperature hotter: in the morning or afternoon? Explain.** afternoon because she wore lighter clothing

Alternative Teaching Strategy

Place the Temperature

Objective To read and model temperature on a thermometer

Materials large blank paper Fahrenheit thermometer that shows ⁻10° to 100°F, marked and labeled at increments of 10s with tick marks for each "one" in between; index cards with a different temperature written on each

• Explain to children that they will be modeling temperatures on a paper thermometer.

• Model for children what they will do. Say: **I'm going to pick a card.** Display one of the cards such as one for 25°F. **My card has 25 degrees Fahrenheit written on it. So I will find the number 25 on the thermometer. There is no 25, so I have to count up from 20.** Indicate 20 on the thermometer. Then count aloud, indicating each tick mark. Say: **20, 21, 22, 23, 24, 25.**

• Give each child an index card with a temperature on it. Have children identify the temperature written on their cards.

• Invite children to show each temperature on the thermometer.

• On each child's turn, discuss activities that one might do at that particular temperature.

Intensive Intervention · IIN74

Compare Weights
Skill ㊳

Objective
To compare objects by weight

Materials
crayon, notebook, balloon, stapler

Vocabulary
heavier
lighter

Pre-Assess

Show children a crayon and a notebook. Ask: **Which object is heavier?** the notebook **Which object is lighter?** the crayon **How do you know?** Possible answer: If you hold both of them in your hands, the notebook feels heavier than the crayon. Next ask children to compare the weights of a real car and a toy car. **Which is heavier?** real car Ask: **Which is lighter?** toy car **How can you know this without holding them?** A real car is too big and heavy to hold but you can hold a toy car so you know the real car is heavier.

Common Misconception

- Children may believe that the bigger object in a pair is always heavier.

- To correct this, point out the example of a blown-up balloon and a stapler. Explain that even though a balloon is bigger than a stapler, it is filled with air, which is lighter. So the balloon is lighter than the stapler. Tell children when they compare weights to imagine how heavy each object would feel if they were holding it.

Learn the Math

Have a balloon and a stapler for children to hold to compare as you guide them through student page **IIN75**.

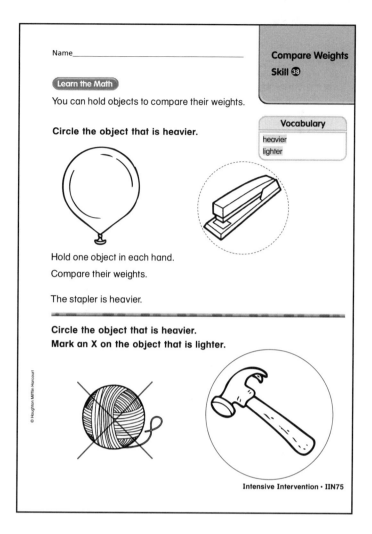

Learn the Math

You can hold objects to compare their weights.

Circle the object that is heavier.

Vocabulary
heavier
lighter

Hold one object in each hand.
Compare their weights.

The stapler is heavier.

Circle the object that is heavier.
Mark an X on the object that is lighter.

© Houghton Mifflin Harcourt

Intensive Intervention · IIN75

Talk Math

- Hold the balloon and the stapler for children to see. **Which of these do you think is heavier?** Answers may vary.

Pass the balloon and stapler so each child has an opportunity to hold them together. As you do this, discuss with children their ideas about which is heavier and have them explain why.

- **Which is heavier: the balloon or the stapler?** the stapler

Have children look at the pictures at the bottom of the page. Then work through the second problem with them.

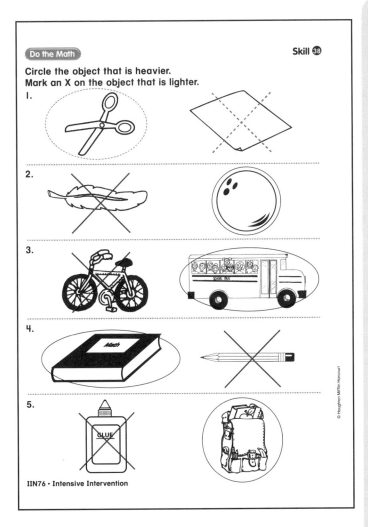

Do the Math

Circle the object that is heavier.
Mark an X on the object that is lighter.

1.
2.
3.
4.
5.

IIN76 · Intensive Intervention

© Houghton Mifflin Harcourt

Alternative Teaching Strategy

Lighter and Heavier

Objective To compare objects by weight

Materials balance, classroom objects that will fit on the balance

- Explain to children that they will compare the weights of two objects by using a balance. Show children the balance. Explain that when two objects have the same weight, the balance will be even. If one object is heavier than the other, the side with the heavier object will go down and the side with the lighter object will go up.

- Demonstrate by placing a paper clip on each side. Ask: **Are the sides even?** yes **What do you think will happen if I put paper clips on one side and not the other?** The side with paper clip will go down and the side with no paper clips will go up.

- Show children a pencil and a marker. Ask: **Which do you think is heavier?** Answers will vary. Explain that sometimes it is difficult to tell which object is heavier unless you use a balance. Place the pencil and marker on opposite sides of the balance. Discuss with children which is heavier and if their predictions were correct.

- Repeat with other classroom objects to explore comparing weights with a balance.

Do the Math

Have children look at student page **IIN76**. Explain to children that they will compare two objects to find which is heavier and which is lighter. They will circle the heavier object and put an X on the lighter object. If the objects are found in the classroom, they may hold the objects to compare.

Talk Math

- **How can you compare the weights of two objects that you cannot hold?** You can imagine you are holding them.

- **How do you know which object is heavier when you are holding two objects?** Possible answer: It feels heavier. It is harder to hold.

Check

Ask children to explain whether their shoe or desk is heavier.

Compare Capacities
Skill ㊴

Objective
To compare the capacities of containers

Materials
containers to compare capacities such as a gallon milk carton, large cooking pot, water bottle, drinking glass, large bucket, teaspoon, cereal bowl, and bucket; dry rice or water

Vocabulary
capacity
less
more

Pre-Assess
Show children a large container such as a gallon milk carton and a smaller container such as a drinking glass. Ask: **Which container holds more?** the milk carton **How do you know it holds more?** It is bigger. It has more space to hold something. **Which container holds less?** the drinking glass Continue with other sets of objects that hold different amounts.

Common Misconception
• Children may confuse capacity and weight.

• To correct this, explain that something that has a greater capacity, may weigh much less than something else with a smaller capacity. Point out that a heavy bowl might weigh more than a milk carton, but a milk carton might hold much more than a bowl.

Learn the Math

You may wish to have the actual objects shown on student page **IIN77** available to demonstrate how to compare capacities. To compare capacities, fill one container with water or dry rice and pour the contents into the second container. Help children see that if there is water or rice left over, then the first container holds

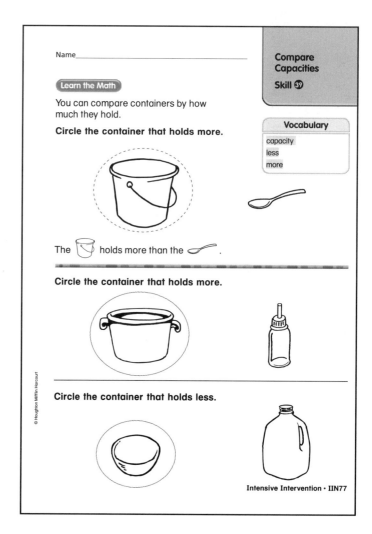

more than the second. If the second container is not full, then it has a greater capacity.

Talk Math

• **How can you tell which of two containers holds more?** Possible answer: I can fill one container with water. Then I can pour it into the other container. If there is some left over, the second container holds more.

• **How do you know a bucket would hold more water than a teaspoon?** A bucket is much bigger than a teaspoon. It has more space to hold something.

Guide children through comparing the capacities of the two containers in each of the problems on student page **IIN77**.

IIN78 · Intensive Intervention

Do the Math

Have children look at student page **IIN78**. Explain that they will compare the capacities of two containers. In Problems 1–4, they will circle the container that holds more. In Problems 5–8, they will circle the container that holds less.

Talk Math

- **How will you decide which container holds more?** It will have more space to hold something.

- **How will you decide which container holds less?** It will have less space to hold something.

Check

Ask: **What does it mean when a container holds less?** It holds less than the other container you are comparing it to.

Alternative Teaching Strategy

Compare and Measure

Objective To compare capacities of containers

Materials large container such as a sand bucket, smaller container such as a bowl, enough dry rice to fill both containers, 1-cup measuring cup

- Explain to children that they will compare the capacities of two containers and then measure the capacities to find out if they were correct.

- Place two containers, such as a bucket and a small bowl, on a table in front of the class. Ask: **Which of these containers do you think will hold more?** the bucket

- Show children the rice and the measuring cup. Ask: **How do you think you could use these two things to test which container holds more?** Possible answer: You could use the cup to fill each container with rice and see which holds more cups.

- Say: **Let's fill the containers with rice using the measuring cup. Help me count how many cups you need to fill the bucket.** Fill the bucket with cups of rice. Count each cup of rice with children and write the total number of cups on the board.

- Say: **Now let's count how many cups you need to fill the bowl.** Fill the bowl with cups of rice. Count each cup with children and write the total number of cups on the board.

- Ask: **Which container holds more?** the bucket **How do you know?** It holds more cups of rice than the bowl.

Objective
To recognize equal parts of a whole

Materials
paper, scissors

Vocabulary
equal parts

Pre-Assess

Draw a square on the board. Draw a vertical line segment through the middle of the square to divide it into equal parts. Then draw a congruent square to the right of the first square. Draw a vertical line segment through the square to divide it into two unequal parts. Point to the first square. Ask: **How many equal parts are there?** two Point to the second square. Ask: **How many equal parts are there?** none Draw a circle and divide it into four equal parts. Ask: **How many equal parts are there?** four

Common Misconception

• Children may have difficulty recognizing equal parts.

• To correct this, explain that when a shape is divided into equal parts, all of the parts are the same size. Demonstrate for children by folding a sheet of paper in halves or in quarters and cutting out each part. Point out that the parts, when placed on top of each other are the same size.

Learn the Math

Point out that the first circle on student page **IIN79** has two equal parts and the second circle has two unequal parts. Explain that equal parts are the same size and unequal parts are not the same size.

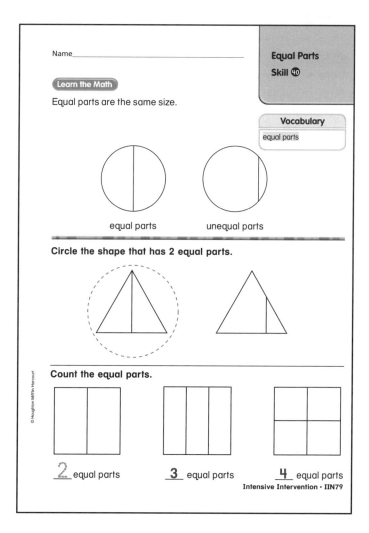

Talk Math

• Direct children to the first triangle. **Does the first triangle have equal parts or unequal parts?** equal **How do you know?** The parts are the same size.

• **Does the second triangle have equal or unequal parts?** unequal **How do you know?** One part is bigger than the other.

Have children trace the circle around the first triangle. Guide children through counting the equal parts in each shape at the bottom of the page. Have them write the number of equal parts for each shape.

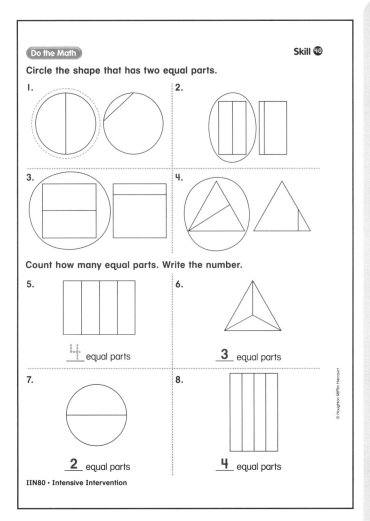

Do the Math

Have children look at student page **IIN80**. Explain that in Problems 1 – 4, they will circle the shape that has equal parts. In Problems 5 – 8, they will count how many equal parts the shape has and write that number.

Talk Math

• **How do you know if two parts are equal?** They are the same size.

• **How do you find how many equal parts a shape has?** I count each part.

Check

Ask: **How are equal parts of a shape alike?** They are all the same size.

Alternative Teaching Strategy

Make Equal and Unequal Parts

Objective To explore equal and unequal parts

Materials paper, scissors, two-dimensional shapes (see Teacher Resource Book), including circles, squares, and rectangles

• Give each child a pair of scissors and a few sheets of paper. Explain that they will be tracing and cutting the shapes to show equal and unequal parts.

• Provide children with shapes such as circles, squares, and rectangles. Have them select a shape to trace. Direct them to cut out the shape, fold it evenly, then cut it into two equal parts.

• Show children how to place the two parts on top of each other. Point out that if they are equal, they are the same size so they will fit perfectly on top of each other. Ask: **Are the two parts equal or unequal?** equal

• Next, have them trace the same shape and cut it out again. This time, direct them to fold it to make two unequal parts. Then cut along the fold. Remind children that unequal parts are different sizes. Ask: **Are the two parts equal or unequal?** unequal **How do you know they are unequal?** I placed the two parts on top of each other and they are not the same size.

• Continue the activity using different shapes and different numbers of equal parts.

Explore Halves
Skill ④

Objective
To identify halves

Vocabulary
one half ($\frac{1}{2}$)

Pre-Assess

Draw two squares of the same size on the board. Divide one square into halves and the other square into thirds. Ask: **Which square shows halves? Explain.** The first square shows halves because it is divided into two equal parts. Ask a volunteer to color $\frac{1}{2}$ of the square. Repeat the activity with other shapes.

Common Misconception

• Children may believe that any shape that is divided into two parts is divided into halves.

• To correct this, point out that halves are two equal parts of a whole. Show children a circle cut into two equal parts. Stack halves on top of each other to show that they are the same size. Then show children a circle cut into two unequal parts. Stack parts on top of each other to show that they are not the same size.

Learn the Math

Guide children through the illustrations and problems on student page **IIN81**.

Talk Math

• Direct children to the circle at the top of the page. **How many parts of the circle are there?** 2 parts **Are the two parts equal?** yes

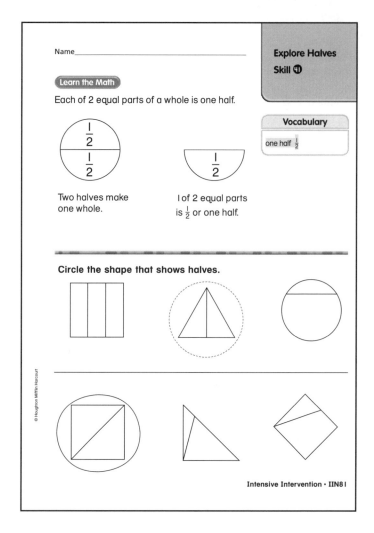

• Direct children to the half circle. **How much of the circle is shown?** one half

Point to each shape in the first row and discuss whether each shape shows halves. Encourage children to explain their thinking. Then have them trace the circle around the triangle. Repeat for the second row.

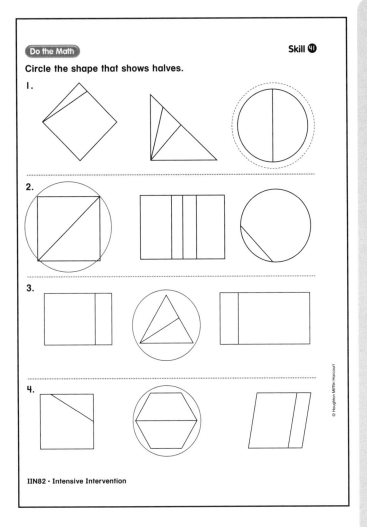

Do the Math

Read the directions aloud on student page **IIN82**. Remind children that halves can only be made with equal parts.

Talk Math

Direct children to Problem 1.

- **Are the two parts of the first shape the same size?** no

- **Now look at the second shape. How many parts are there?** 3 parts

- **Look at the circle. How many parts are there?** 2 parts **Are the two parts the same size?** yes

Direct children to trace the circle around the shape that has two equal parts. Then guide them through Problems 2–4.

Check

Direct students to color one half of each shape they circled in Problems 2–4.

Alternative Teaching Strategy

Can I Have Half?

Objective To explore halves using real-world objects

Materials pieces of string about one meter long

- Separate children into pairs. Give each pair of children a piece of string. Tell them they will use the string to divide objects in the classroom into halves.

- Demonstrate by placing a piece of string over a book to divide it in half. Say: **I am going to use my string to divide the book in half.** Point to one half. **Here is one half.** Point to the other half. **Here is the other half. They are equal.**

- Point out to children that the halves are the same size. Place the string over a pencil. Point out that you cannot divide the pencil into halves by putting the string in the middle because the halves will be different. One will have an eraser and the other will have a point. Show children how to place the string lengthwise down the pencil to divide it into halves.

- Say: **Work with your partner to find five classroom objects to divide in half.**

- After children are finished, have each pair share an object with the class and show how the object is divided into halves.

Explore Thirds and Fourths
Skill ㊷

Objective
To identify thirds and fourths

Materials
fraction strips (see Teacher Resource Book)

Pre-Assess

Draw two circles on the board. Divide the first circle into thirds and the second circle into fourths. Ask: **Which circle shows fourths? Explain.** The second circle shows fourths. It is divided into 4 equal parts. Ask a volunteer to color one fourth of the circle. Ask: **What does the other circle show? Explain.** It shows thirds. It is divided into 3 equal parts. Ask a volunteer to color one third of the circle.

Common Misconception

• Children may believe that any shape that is divided into three parts is divided into thirds and any shape divided into four parts is divided into fourths.

• To correct this, remind children that a shape must be divided into equal parts for it to be divided into thirds or fourths.

Learn the Math

Have children use fraction strips to model each each fraction as you guide them through the examples at the top of student page **IIN83**.

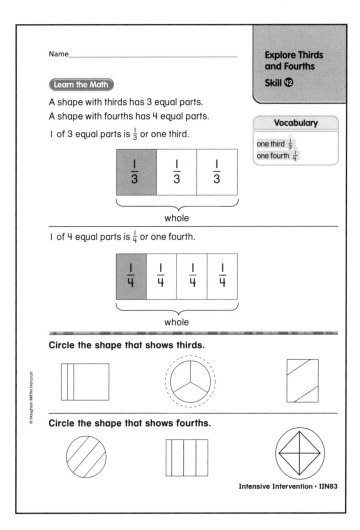

Talk Math

• **How many equal parts is the first fraction strip divided into?** 3 equal parts **How many of the parts are shaded?** 1 part

• **What fraction is the shaded part?** one third

• **How many equal parts is the second strip divided into?** 4 equal parts **How many of the parts are shaded?** 1 part

• **What fraction is the shaded part?** one fourth

Guide children as they complete the problems at the bottom of the page. Remind them that a shape shows thirds or fourths only when the parts are the same size.

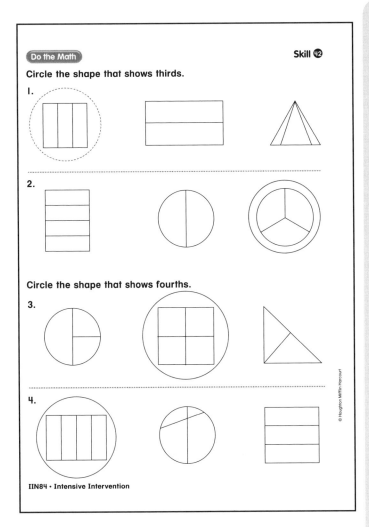

Do the Math

Have children look at student page **IIN84**. Explain to them that in the first two problems, they will circle the shape that shows thirds. In the next two problems, they will circle the shape that shows fourths. Remind them that the parts must be equal.

Talk Math

• **How can you tell if a shape shows thirds?** It has three equal parts.

• **How can you tell if a shape shows fourths?** It has four parts that are equal in size.

Check

Have children color one third of each shape they circled in Problems 1 and 2. Have them color one fourth of each shape they circled in Problems 3 and 4.

Alternative Teaching Strategy

Making Models of Thirds and Fourths

Objective To explore thirds and fourths with fraction strips

Materials blank strips of paper to use as fraction strips, rulers

• Give each child two blank fraction strips of the same length.

• Tell them to take one fraction strip and fold it in half. Say: **Unfold the strip and draw a line on the fold.** Ask: **How many equal parts is the strip divided into?** 2 equal parts

• Say: **Fold the strip again and then again. This time when you unfold it, you should see two more fold lines. Draw a line on each fold line.**

• Ask: **How many equal parts is the strip divided into?** 4 equal parts **Does the strip show halves, thirds, or fourths?** fourths Say: **Color one fourth.**

• Say: **Now take the other fraction strip and fold it to show thirds.** Demonstrate how to fold the strip into thirds. Assist students who are having trouble.

• Say: **Unfold the fraction strip and draw lines on the fold lines. Color one third.**

Objective
To use models to show and record 2-digit numbers to 50

Materials
base-ten blocks

Vocabulary
tens
ones

Pre-Assess

Model 35 using base-ten blocks. Ask: **How many tens are there?** 3 tens Write 3 on the board. **How many ones are there?** 5 ones Write 5 to the right of 3 on the board. **What number do the base-ten blocks show?** 35 Give each child 5 tens and 9 ones. Repeat with other numbers up to 50. Each time, have children show the number using their base-ten blocks and then record the number on a sheet of paper.

Common Misconception

- Children may write the ones in the tens place and the tens in the ones place.

- To correct this, draw a place-value chart on the board. Label the left column "tens" and the right column "ones." Tell children that each time they write a 2-digit number they have to write a digit for tens and a digit for ones. Remind them to write the tens digit first and the ones digit to the right of the tens digit.

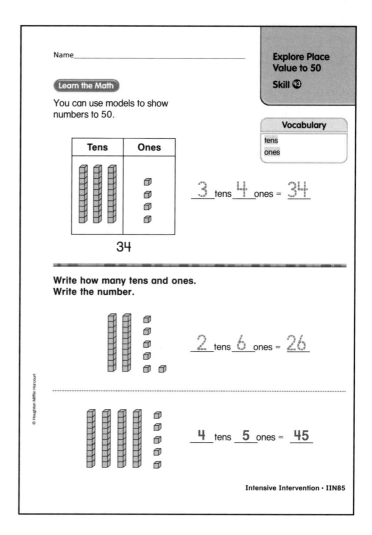

Learn the Math

Give children base-ten blocks to model the numbers on student page **IIN85**. You may wish to remind them that 1 ten is equal to 10 ones.

Talk Math

- Direct children to the first problem. **How many tens are there?** 3 tens

- **How many ones are there?** 4 ones

- **What is the number?** 34

Ask similar questions for the next two problems. Have children write the number of tens and ones and then write the number.

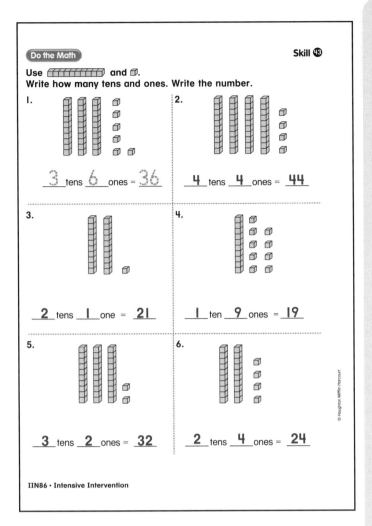

Do the Math   Skill 43

Use ▭ and ▢.
Write how many tens and ones. Write the number.

1.
___3___ tens ___6___ ones = ___36___

2.
___4___ tens ___4___ ones = ___44___

3.
___2___ tens ___1___ one = ___21___

4.
___1___ ten ___9___ ones = ___19___

5.
___3___ tens ___2___ ones = ___32___

6.
___2___ tens ___4___ ones = ___24___

IIN86 · Intensive Intervention

© Houghton Mifflin Harcourt

Alternative Teaching Strategy

Up To 50

Objective To use models to show and record 2-digit numbers to 50

Materials connecting cubes

- Separate children into pairs. Give each pair of children 50 connecting cubes.

- Ask children to connect 10 cubes to make 1 cube train. Then have them make a second cube train of 10 connecting cubes. Tell children to lay the two cube trains on their desks. Tell them that each cube train stands for 1 ten.

- Next, have children count out 7 connecting cubes and put them to the right of the two cube trains.

- Ask: **How many tens are there?** 2 tens **How many ones are there?** 7 ones **What is the number shown by the model?** 27

- Have pairs work together to model and record other numbers to 50.

Do the Math

Have children look at student page **IIN86**. Explain to them that they will use their base-ten blocks to model each number. Then they will record how many tens, how many ones, and the number shown in each model.

Talk Math

Guide children through Problem 1 with the following questions.

- **How many tens are shown?** 3 tens

- **How many ones are shown?** 6 ones

- **What number is shown by the base-ten blocks?** 36

Repeat with similar questions for Problems 2–6.

Check

Ask: **For Problem 1, James wrote 63. What mistake did James make?** He put the 6 in the tens place and the 3 in the ones place.

Explore Place Value to 100
Skill ④④

Objective
To use models to show and record 2-digit numbers to 100

Materials
base-ten blocks

Vocabulary
tens
ones

Pre-Assess
Model 87 using base-ten blocks. Ask: **How many tens are there?** 8 Write 8 on the board. **How many ones are there?** 7 Write 7 to the right of 8 on the board. Ask: **What number do the base-ten blocks show?** 87 Give each child 9 tens blocks and 9 ones blocks. Repeat with other numbers up to 100. Each time, have children show the number using their base-ten blocks and then record the number on a piece of paper.

Common Misconception

• Children may incorrectly count the number of tens.

• To correct this, have children mark each ten as they count it. Then have them check the number of tens by counting them a second time.

Learn the Math

Give children base-ten blocks to model the numbers on student page **IIN87**. You may wish to remind them that one tens block is equal to 10 ones.

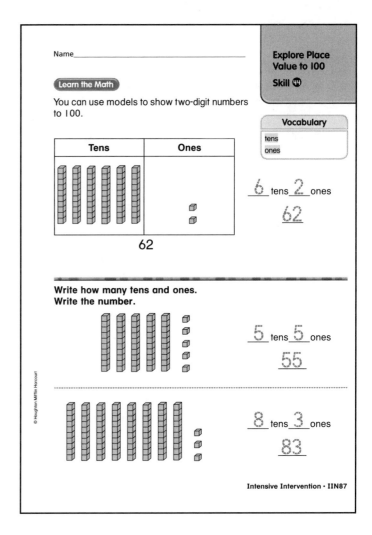

Talk Math

Direct students to the first problem.

• **Look at the tens column on the chart. Count the tens with me, 1 ten, 2 tens, 3 tens, 4 tens, 5 tens, 6 tens. How many tens are there?** 6 **What is the value of the 6?** 60

• **Look at the ones column. How many ones are there?** 2 **What is the value of the 2?** 2

• **What number is shown in the model?** 62

Repeat with the next two problems. Have children trace the dashed answers in both problems.

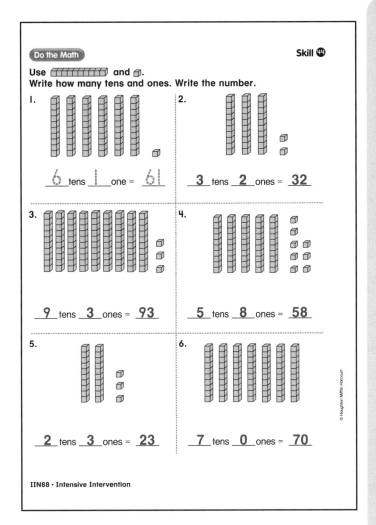

Do the Math

Have children look at student page **IIN88**. Give children base-ten blocks to model each number. Point out that the numbers may go up to 100. Tell them to count the tens and ones to find the number shown by the base-ten blocks.

Talk Math

Guide children through Problem 1.

- **How many tens are there?** 6 tens
- **How many ones are there?** 1 one
- **What number is shown in the model?** 61

Check

Have students model 99 with base-ten blocks. Then ask them to add one to 99. Tell them that after 99, comes 100. Ask: **How many tens are there in 100?** 10 tens

Alternative Teaching Strategy

Up To 100

Objective To use models to show and record 2-digit numbers to 100

Materials dry beans, strips of paper, and glue

- Divide children into groups of four. Tell students that they will be modeling numbers up to 100 using dry beans.

- On the board, write the number 52. Say: **One way to count 52 beans is to make groups of ten.**

- Encourage children to share the task of counting the beans by having each group member count a group of ten. Then they can share in the task of counting the last ten. Remind them to add the ones.

- Direct children to glue their groups of ten to a strip of paper. **How many tens are there?** 5 tens **How many ones are there?** 2 ones

- Repeat with other numbers up to 100. Be sure to include some numbers less than 50.

Objective
To use models to compare numbers to 50

Materials
base-ten blocks

Pre-Assess

Model 18 using base ten blocks. Ask: **What number did I show?** 18 Model 12 using base ten blocks. Ask: **What number did I show?** 12 **Which is the greater number, 18 or 12?** 18 Model 24 using base ten blocks. Ask: **What number did I show?** 24 Model 28 using base ten blocks. Ask: **What number did I show?** 28 **Which is the lesser number, 24 or 28?** 24 Show children two models of 14 using base ten blocks. Ask: **Are these numbers equal?** yes **How do you know?** They have the same number of tens and ones.

Common Misconception

- Children may assume that if one number has more ones than the number it is being compared to, it must be greater.

- To correct this, point out to children that they need to compare the tens first. If the tens are not the same, the number with more tens is the greater number. If the tens are the same, then they should compare the ones. Then the number with more ones is the greater number.

Learn the Math

Provide base-ten blocks to model the problems on student page IIN89.

Talk Math

- Direct children to the first problem. **How many tens are there in the first number?** 3 **How many ones?** 4 **What number is shown?** 34

- **How many tens are there in the second number?** 2 **How many ones?** 5 **What number is shown?** 25

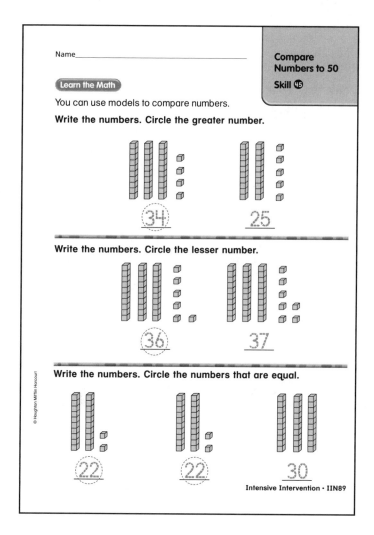

Name_____

Compare Numbers to 50
Skill ④⑤

Learn the Math

You can use models to compare numbers.

Write the numbers. Circle the greater number.

(34) 25

Write the numbers. Circle the lesser number.

(36) 37

Write the numbers. Circle the numbers that are equal.

(22) (22) 30

Intensive Intervention · IIN89

- **Compare the tens. Which is greater, 3 tens or 2 tens?** 3 tens **Which number is greater, 34 or 25?** 34

Have children trace the numbers and the circle around the greater number.

Tell children to write the numbers shown in the second problem.

- **Which number has fewer tens?** They have the same number of tens.

- **Compare the ones. Which number has fewer ones?** 36

- **Which is the lesser number?** 36

For the last problem, have children compare the first two numbers, and then the third number. Have children trace over the dashed answers.

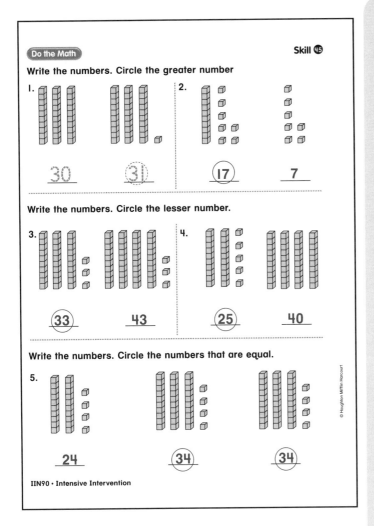

Alternative Teaching Strategy

Comparing with a Number Line

Objective To compare numbers to 50 using a number line

Materials blank number lines that range from 0 to 50

- Give each child a number line. Explain to children that as you move to the right on a number line the numbers increase and as you move to the left the numbers decrease.

- Write 16 and 21 on the board. Say: **Use your number line to tell which number is greater. First, find 16 and circle it. Then find 21 and circle it.** Ask: **Which number is farther to the right, 16 or 21?** 21 **Which is the greater number, 16 or 21?** 21

- Continue the activity with different pairs of numbers. Alternate asking children to tell which number is greater and which number is lesser.

Do the Math

Have children look at student page **IIN90**. Tell them that they will write each number and then either circle the greater number, the lesser number, or the numbers that are equal. Tell them to compare the tens first and then the ones.

Talk Math

- **Why do you need to compare the tens first?** the tens have greater value than the ones

- **How do you know if two numbers are equal?** They have the same number of tens and ones.

Check

Ask: **Peri says that 17 is greater than 26. Is she correct? Explain.** Possible answer: No, 17 is less than 26. 26 has 2 tens and 6 ones. 17 has 1 ten and 7 ones. Since 26 has more tens, it is the greater number.

Objective
To use a number line to determine before, between, and after

Vocabulary
after
before
between

Pre-Assess
Draw a number line that ranges from 50–60 on the board. Ask: **What number is just before 53?** 52 Draw a square around the number 52. **What number is just after 53?** 54 Draw a square around the number 54. **What number is between 52 and 54?** 53

Common Misconception
- Children may confuse before and after on a number line.

- To correct this, point out that a number which comes just after another number is the first number to the right of that number on the number line. A number which comes just before another number is the first number to the left of that number on the number line.

Learn the Math

Have children point to each number on the number line on student page **IIN91** as you discuss finding numbers just before, between, and just after.

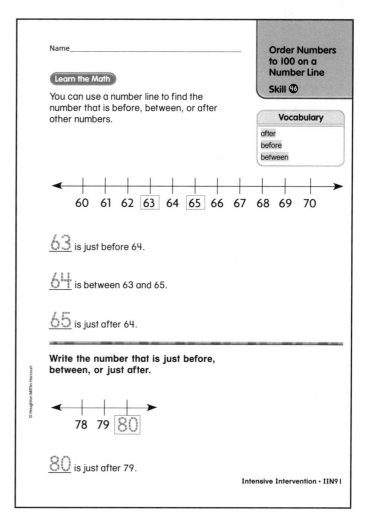

Talk Math

- **Find 64. What number comes just before 64?** 63 Have children trace the number 63.

- **What number is between 63 and 65?** 64 Have children trace the number 64.

- **What number is just after 64?** 65 Have children trace the number 65.

- **What number comes just before 63?** 62

- **What number comes just after 65?** 66

Direct children to the problem at the bottom of the page. Ask them what number comes just after 79. Have them trace the number 80.

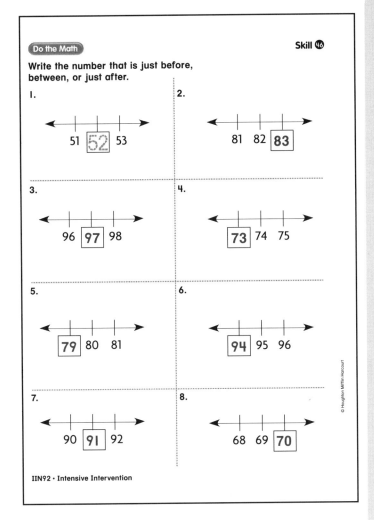

Do the Math

Skill 46

Write the number that is just before, between, or just after.

1.

51 [52] 53

2.

81 82 [83]

3.

96 [97] 98

4.

[73] 74 75

5.

[79] 80 81

6.

[94] 95 96

7.

90 [91] 92

8.

68 69 [70]

IIN92 · Intensive Intervention

© Houghton Mifflin Harcourt

Alternative Teaching Strategy

Greater Numbers, Greater Number Lines

Objective To use a number line to determine before, between, and after

Materials sidewalk chalk or masking tape

- If possible, create a number line from 80 to 90 on the sidewalk using chalk. If not, you can create a number line on the classroom floor using masking tape.

- Ask a volunteer to stand on 83 on the number line. Ask: **What number is just before 83?** 82 **What number is just after 83?** 84

- Have another volunteer stand on 85 on the number line. Ask: **What number is between 83 and 85?** 84

- Have another volunteer stand on 88 on the number line. Ask: **What number is just before 88?** 87 **What number is just after 88?** 89 **What number is between 88 and 90?** 89

- Continue the activity with different numbers until each child has had the opportunity to stand on the number line.

Do the Math

Have children look at student page **IIN92**. Point out that they need to find the number that is missing. Explain that the number is either just before another number, just after another number, or between two numbers.

Talk Math

- **Do you move to the right or left to find a number that is just before another number?** to the left

- **Do you move to the right or left to find a number that is just after another number?** to the right

Check

Tell children to choose a number between 90 and 100. Have them write the number that comes just before their number and the number that comes just after it.

© Houghton Mifflin Harcourt

Objective

To model addition with 1-digit and 2-digit numbers with and without regrouping

Materials

base-ten blocks

Pre-Assess

Give children base-ten models. Say: **Add 23 and 5.** Ask: **What number will you show first?** 23 **How can you show 23 with base-ten blocks?** 2 tens blocks and 3 ones blocks **What blocks will you use to add 5?** 5 ones blocks **Can you regroup 10 ones as a ten?** no Ask: **How many tens and ones do you have in all?** 2 tens 8 ones Ask: **What is the sum of 23 and 5?** 28

Repeat the activity by having children add 34 and 9. Ask similar questions.

Common Misconception

• Children may forget to add the ten after they have regrouped ones.

• To correct this, encourage children to draw a box over the tens place to help them remember whether they need to add a ten.

Learn the Math

Have children use base-ten blocks to model the problem on student page **IIN93**.

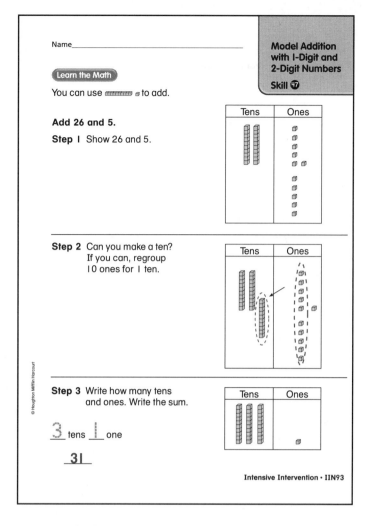

Talk Math

• **Look at the ones in the first step. Can you make a ten from the ones?** yes

• **Where do you put the ten after you regroup?** in the tens column

• **How many ones are left after you make a ten?** 1

• **How many tens are there now?** 3

• **What is the sum of 26 and 5?** 31

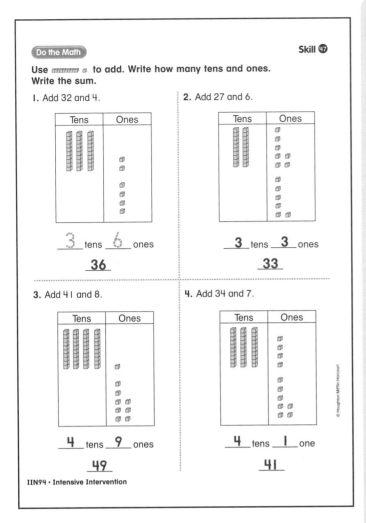

Do the Math

Skill ⑰

Use ▭ ▫ to add. Write how many tens and ones. Write the sum.

1. Add 32 and 4.

Tens	Ones

___3___ tens ___6___ ones

___36___

2. Add 27 and 6.

Tens	Ones

___3___ tens ___3___ ones

___33___

3. Add 41 and 8.

Tens	Ones

___4___ tens ___9___ ones

___49___

4. Add 34 and 7.

Tens	Ones

___4___ tens ___1___ one

___41___

IIN94 · Intensive Intervention

© Houghton Mifflin Harcourt

Do the Math

Have children look at student page **IIN94**. Explain that they need to find the sum in each problem. Point out that not all problems will require them to regroup. They should first decide if they can make a ten to know whether they need to regroup.

Talk Math

- **How do you decide if you need to regroup?** Count the number of ones blocks. If there are 10 or more, then I need to regroup.

- **What do you do if the sum of the ones is ten or more?** regroup 10 ones as 1 ten

Check

Ask: **Do you need to regroup when you add 15 and 8? Explain.** Yes, the sum of the ones is 13, which is more than 10.

Alternative Teaching Strategy

Add It All Together

Objective To model addition of 1-digit and 2-digit numbers

Materials connecting cubes

- Give each child 40 connecting cubes.

- Review with children that 10 ones makes 1 ten.

- Guide children to model 12 + 6 in a workmat.

TENS	ONES

12
+ 6

- Start with 12. Ask: **How many tens are in 12?** 1 Direct children to make a ten with their connecting cubes and to place the ten in the tens column. **How many ones are in 12?** 2 Direct children to place 2 ones in the ones column.

- Say: **Now we need to add 6.** Ask: **How many tens are in 6?** 0 **How many ones are in 6?** 6 Direct children to place 6 ones in the ones column under the 2 ones.

- Write 12 + 6 next to the place-value chart. Say: **This shows 12 plus 6.**

- Say: **Add the ones. How many are there in all?** 8 **Now add the tens.** Ask: **How many tens are there in all?** 1 **So what is the sum of 12 and 6?** 18

- Repeat the activity, however this time, write an addition problem in which children have to regroup to solve the problem.

- Continue the activity with different problems. You may wish to pair children together and have them think of addition problems for each other to solve.

Intensive Intervention · IIN94

Objective
To use mental math to add tens

Materials
base-ten blocks

Pre-Assess

Write the following problem on the board:

$$\begin{array}{r} 30 \\ + 40 \\ \hline \end{array}$$

Say: **Use mental math.** Ask: **What is the sum of 30 and 40?** 70 Erase the problem and write 40 + 20 on the board. Say: **Use mental math.** Ask: **What is the sum of 40 and 20?** 60 Continue with other addition problems to determine whether children can add tens.

Common Misconception

• Children may have difficulty relating basic facts to adding tens.

• To correct this, use tens blocks to model tens. Have children count the number of tens blocks, then count by tens to find the value of the blocks. Record the number of blocks and then the value of the blocks on the board, such as, 3 tens blocks and 30 or 4 tens blocks and 40. Help children see the pattern: the number of blocks is the same as the number of tens. Point out that if you add 3 tens blocks and 1 tens block you will have 4 tens blocks or 40.

Learn the Math

Guide children through the problems on student page **IIN95**.

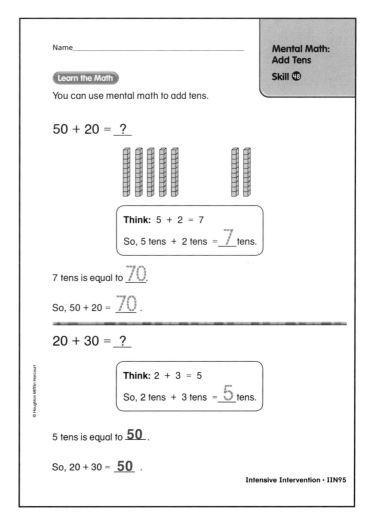

Learn the Math

You can use mental math to add tens.

50 + 20 = _?_

Think: 5 + 2 = 7

So, 5 tens + 2 tens = _7_ tens.

7 tens is equal to _70_.

So, 50 + 20 = _70_.

20 + 30 = _?_

Think: 2 + 3 = 5

So, 2 tens + 3 tens = _5_ tens.

5 tens is equal to _50_.

So, 20 + 30 = _50_.

© Houghton Mifflin Harcourt

Intensive Intervention • IIN95

Talk Math

• **How many tens blocks do you use to show 50?** 5

• **How many tens blocks do you use to show 20?** 2

• **How does knowing 5 + 2 = 7 help you find 50 + 20?** Possible answer: Since 5 and 2 are the number of tens blocks I am adding, I know that I will have 7 tens blocks. If I have 7 tens blocks I know that is 70.

Point out to children that when they add tens, the digit in the ones place is always zero, so they can just add the digits in the tens place. Guide children through the second problem on the page.

Do the Math

Skill 48

Add.

1. $4 + 3 = 7$

$\underline{4}$ tens $+ \underline{3}$ tens $= \underline{7}$ tens

$40 + 30 = 70$

2. $5 + 4 = 9$

5 tens $+ 4$ tens $= \underline{9}$ tens

$50 + 40 = 90$

3. $2 + 6 = 8$

2 tens $+ 6$ tens $= \underline{8}$ tens

$20 + 60 = 80$

4. $\begin{array}{r} 30 \\ + 60 \\ \hline 90 \end{array}$	5. $\begin{array}{r} 10 \\ + 10 \\ \hline 20 \end{array}$	6. $\begin{array}{r} 20 \\ + 70 \\ \hline 90 \end{array}$
7. $\begin{array}{r} 40 \\ + 40 \\ \hline 80 \end{array}$	8. $\begin{array}{r} 60 \\ + 10 \\ \hline 70 \end{array}$	9. $\begin{array}{r} 10 \\ + 20 \\ \hline 30 \end{array}$

IIN96 · Intensive Intervention

Alternative Teaching Strategy

Toss and Add

Objective To use mental math to add tens

Materials number cubes

- Give each pair of children two number cubes.

- Say: **Each partner should take one of the number cubes. Toss your number cube. Now use the numbers on the number cube to write an addition sentence for adding tens. For example, if I tossed 2 and 3, I would write 2 tens + 3 tens or 20 + 30.**

- Say: **Use mental math to complete the addition sentence. I know that 2 + 3 = 5, so I can use mental math to write 20 + 30 = 50.**

- Have children toss the number cubes to generate 5 or more addition sentences.

- Have pairs of children trade papers with other pairs and use mental math to check the addition.

Do the Math

Have children look at student page **IIN96**. Point out that in the first three problems they will first complete the basic fact and add the number of tens to help find the final sum. In the remaining problems, they can do these steps mentally.

Talk Math

- **How many ones are in 1 ten?** 10 ones

- **Why do you add only the tens digits when you add tens?** because the ones digit is always zero

Check

Ask: **What basic fact will help you find 50 + 30?**
$5 + 3 = 8$

Objective
To model subtraction with 2-digit and 1-digit numbers with and without regrouping

Materials
base-ten blocks

Pre-Assess

Give children base-ten blocks. Say: **Subtract 7 from 21.** Ask: **What number will you show with the base-ten blocks?** 21 **How can you show 21 with base-ten blocks?** 2 tens blocks and 1 ones block **Can you subtract 7 ones? If not, explain why.** No, there are not enough ones. **What do you do when there are not enough ones to subtract?** Regroup 1 ten as 10 ones. Have children regroup the blocks then subtract. Ask: **How many tens and ones do you have in all?** 1 ten 4 ones Ask: **What is the difference?** 14

Repeat the activity by having children subtract 4 from 32. Ask similar questions.

Common Misconception

- Children may always want to subtract the lesser number of ones from the greater number of ones.

- To correct this, write 23 – 8 on the board. Say: **23 take away 8. You must start with 23 and take away 8 ones. If there are not enough ones, then you have to regroup 1 ten as 10 ones.** Repeat with several examples.

Learn the Math

Give children base-ten blocks to model the problem on student page **IIN97**.

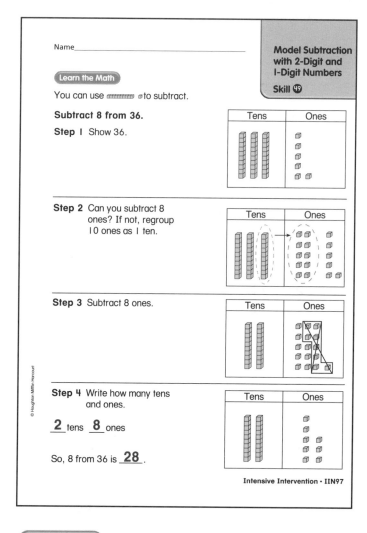

Learn the Math

You can use [blocks] to subtract.

Subtract 8 from 36.

Step 1 Show 36.

Tens	Ones

Step 2 Can you subtract 8 ones? If not, regroup 10 ones as 1 ten.

Tens	Ones

Step 3 Subtract 8 ones.

Tens	Ones

Step 4 Write how many tens and ones.

2 tens **8** ones

So, 8 from 36 is **28**.

Tens	Ones

Intensive Intervention · IIN97

Talk Math

- **Look at the ones blocks in Step 1. Can you subtract 8 ones? Explain.** No, there are only 6 ones.

- **What do you need to do since there are not enough ones to subtract?** I need to regroup 1 ten as 10 ones.

- **How many ones are there after you regroup?** 16 ones

- **Subtract 8 ones from 16 ones. How many ones are left?** 8 ones

- **How many tens are there?** 2 tens

- **What is 36 – 8?** 28

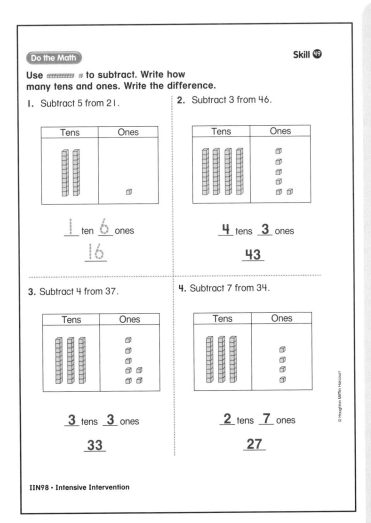

Do the Math

Have children look at student page **IIN98**. Explain that they will find the difference for each problem. Point out that not all problems will require them to regroup.

Talk Math

• **How do you decide if you need to regroup?** Look at the ones. If there are less ones than the number being subtracted, then I need to regroup.

• **What do you do if you do not have enough ones to subtract?** I need to regroup 1 ten as 10 ones.

Check

Ask: **Do you need to regroup when you subtract 7 from 26? Explain.** Yes, you cannot subtract 7 ones from 6 ones.

Alternative Teaching Strategy

Number Line Subtraction

Objective To model subtraction of 1-digit numbers from 2-digit numbers on a number line

Materials number lines that range from 0 to 50 (see Teacher Resource Book), base-ten blocks

• Give each child a number line that ranges from 0 to 50.

• Ask: **When you subtract on a number line, which direction do you move, right or left?** left

• Write 43 – 8 on the board. Say: **Find 43 on your number line.** Ask: **How many jumps to the left will you move to subtract 8?** 8 jumps

• Say: **Move 8 jumps to the left of 43.** Ask: **What is 43 – 8?** 35

• Write 4 more problems on the board for children to solve using the number line.

• Next, write 4 more problems on the board and have children use base-ten blocks to solve them. After they have written their answers, have them use their number lines to check their answers.

Mental Math: Subtract Tens
Skill 50

Objective
To use mental math to subtract tens

Materials
base-ten blocks

Pre-Assess

Write 60 – 30 on the board. Say: **Use mental math.** Ask: **What is 60 – 30?** 30 Write 40 – 20 on the board. Say: **Use mental math.** Ask: **What is 40 – 20?** 20 Continue with other subtraction problems to determine whether children can subtract tens.

Common Misconception

- Children may not understand that they do not need to subtract the ones digits to subtract tens.

- To correct this, write 50 – 10 on the board. Point out that a zero is in the ones place in both numbers. Explain that zero minus zero is always zero. Tell children this means that whenever they subtract a ten from a ten, the ones digit will always be zero.

Learn the Math

Guide children through the examples on student page **IIN99**. Have children use base-ten blocks to model the first example.

Learn the Math

You can use mental math to subtract tens.

$60 - 40 = \underline{\ ?\ }$

> Think: $6 - 4 = 2$
> So, 6 tens – 4 tens = $\underline{2}$ tens.

2 tens is equal to $\underline{20}$.

So, $60 - 40 = \underline{20}$.

$40 - 10 = \underline{\ ?\ }$

> Think: $4 - 1 = 3$
> So, 4 tens – 1 ten = $\underline{3}$ tens.

3 tens is equal to $\underline{30}$.

So, $40 - 10 = \underline{30}$.

Intensive Intervention · IIN99

Talk Math

- **How many tens are there in 60?** 6 tens

- **How many tens are there in 40?** 4 tens

- **What is 6 tens – 4 tens?** 2 tens

- **What is the value of 2 tens?** 20

Guide children through the second problem on the page. Remind children that when they subtract tens, there is always a zero in the ones place. Have them write the number of tens and write the difference.

Do the Math

Subtract.

1. $7 - 5 = \underline{2}$

    $7 \text{ tens} - 5 \text{ tens} = \underline{2} \text{ tens}$

    $70 - 50 = \underline{20}$

2. $8 - 4 = \underline{4}$

    $8 \text{ tens} - 4 \text{ tens} = \underline{4} \text{ tens}$

    $80 - 40 = \underline{40}$

3. $5 - 3 = \underline{2}$

    $5 \text{ tens} - 3 \text{ tens} = \underline{2} \text{ tens}$

    $50 - 30 = \underline{20}$

4.	5.	6.
$\begin{array}{r} 90 \\ -50 \\ \hline 40 \end{array}$	$\begin{array}{r} 70 \\ -30 \\ \hline 40 \end{array}$	$\begin{array}{r} 40 \\ -30 \\ \hline 10 \end{array}$

7.	8.	9.
$\begin{array}{r} 50 \\ -10 \\ \hline 40 \end{array}$	$\begin{array}{r} 80 \\ -50 \\ \hline 30 \end{array}$	$\begin{array}{r} 90 \\ -70 \\ \hline 20 \end{array}$

IIN100 · Intensive Intervention

Do the Math

Have children look at student page **IIN100**. Point out that in the first three problems, they will complete the basic fact and subtract the number of tens to help them find the final difference. In the remaining problems, they can do these steps mentally. You might suggest that children use base-ten blocks to check their answers.

Talk Math

• **10 ones are equal to how many tens?** 1 ten

• **Why can you subtract only the tens digits when you subtract tens?** because the ones digit is always zero

Check

Ask: **What basic fact can help you find 80 – 20?**
$8 - 2 = 6$

Alternative Teaching Strategy

Connect to Mental Math

Objective To use mental math to subtract tens

Materials tape, paper, index cards

• Explain to children that when they subtract tens, they only need to subtract the digits in the tens place and that the digit in the ones place will always be zero.

• Write the following problems on the board as shown:

$\begin{array}{r} 40 \\ -20 \\ \hline 0 \end{array}$	$\begin{array}{r} 50 \\ -10 \\ \hline 0 \end{array}$	$\begin{array}{r} 30 \\ -10 \\ \hline 0 \end{array}$	$\begin{array}{r} 70 \\ -40 \\ \hline 0 \end{array}$

• Cover the zeros in each difference by taping a piece of paper over them. Point to the 4 and 2 in the first problem. Ask: **How many tens are in 40?** 4 **How many tens are in 20?** 2

• Ask: **What is 4 minus 2?** 2 Write 2 in the tens place.

• Move on to the next problem. Ask: **What is 5 minus 1?** 4 Write 4 in the tens place.

• Move on the next problem. Ask: **What is 3 minus 1?** 2 Write 2 in the tens place.

• Move on the next problem. Ask: **What is 7 minus 4?** 3 Write 3 in the tens place.

• Direct students back to the first problem. Ask: **What number is hidden behind the paper?** 0 **How do you know?** Zero minus zero is always zero. Remove the sheet of paper. Repeat for each problem.

• Distribute index cards to students. Direct students to write subtraction problems that involve subtracting tens on one side of a card, and the answers on the other side.

• Direct students to work with a partner to practice using mental math to solve the problems on the index cards.

Skip-Count by Twos and Fives
Skill ⑤

Objective
To skip-count by twos and fives

Materials
counters

Pre-Assess

Draw a group of two triangles on the board. Ask: **How many triangles are in the group?** 2 Next to it, draw two more groups of two triangles. Ask: **How many triangles are in each of these groups?** 2 Say: **Skip-count by twos to find how many triangles there are in all.** Ask: **How many triangles are there in all?** 6 Next, draw a group of five dots. Ask: **How many dots are in the group?** 5 Draw three more groups of five dots. Say: **Skip-count by fives to find how many dots there are in all.** Ask: **How many dots are there in all?** 20 Continue with other examples to determine if children are able to skip-count by twos and fives.

Common Misconception

• Children may not understand that they can only skip-count when counting equal groups.

• To correct this, use counters. Divide the counters into equal groups of two or five and others into unequal groups. Have children determine which groups they can skip-count by guiding them to see which groups are equal and which groups are not.

Learn the Math

Review counting by twos from 0 to 20 with children, and by fives from 0 to 50. Then help children work through the first problem on student page **IIN101**.

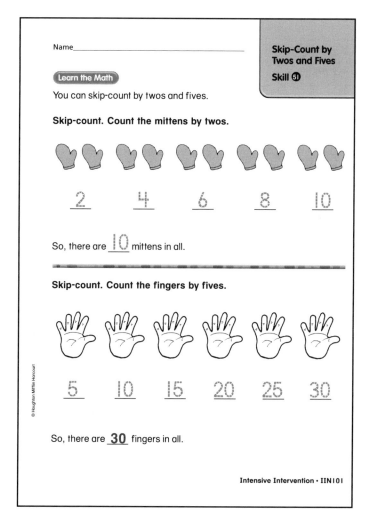

Talk Math

• **How many mittens are in each group?** 2

• **Skip-count by twos to find how many mittens there are in all. What number will you say first?** 2

• **Skip-count. Put your finger on each group of mittens as we count: 2, 4, 6, 8, 10. How many mittens are there in all?** 10

Have children trace over the number of mittens as they count. Repeat for skip-counting the fingers by fives in the second problem.

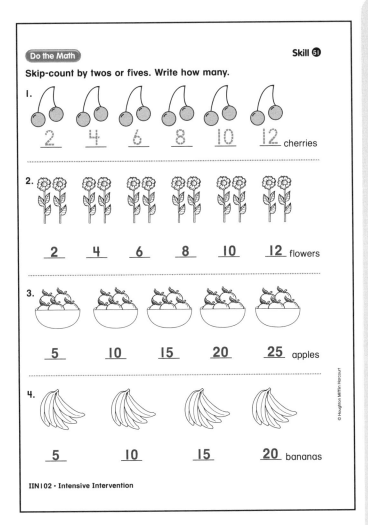

Do the Math

Skip-count by twos or fives. Write how many.

1. 2 4 6 8 10 12 cherries

2. 2 4 6 8 10 12 flowers

3. 5 10 15 20 25 apples

4. 5 10 15 20 bananas

IIN102 · Intensive Intervention

© Houghton Mifflin Harcourt

Skill 51

Do the Math

Have children skip-count by twos and fives on student page **IIN102**. Suggest that they point to each group as they skip-count.

Talk Math

• **How will you know if you should skip-count by twos or fives?** If there are 2 in each group, I will skip-count by twos. If there are 5 in each group, I will skip-count by fives.

• **Why might it be easier to skip-count than to count each number?** It is faster to skip-count because you do not have to say each number.

Check

Ask: **If there are 4 groups of shoes and 2 shoes in each group, how can you skip-count to find how many shoes there are in all?** You can skip-count by twos 4 times: 2, 4, 6, 8. There are 8 shoes in all.

Alternative Teaching Strategy

Skip-count on a Number Line

Objective To skip-count by twos and fives

Materials blank number lines that range from 0 to 40

• Give each child a number line. Explain that they will use the number lines to skip-count by twos and fives.

• Ask: **What is the first number you count when you skip-count by twos?** 2 Ask: **How many spaces will you move on the number line each time you skip-count?** 2

• Say: **Skip-count by twos to 20 on the number line. Circle each number you say.** 2, 4, 6, 8, 10, 12, 14, 16, 18, 20

• Say: **Now skip-count by fives.** Ask: **What is the first number you say when you skip-count by fives?** 5 Ask: **How many spaces will you move on the number line each time you skip-count?** 5

• Say: **Skip-count by fives to 40. Circle each number you say.** 5, 10, 15, 20, 25, 30, 35, 40

Skip-Count on a Hundred Chart
Skill 52

Objective
To skip-count by twos, fives, and tens on a hundred chart

Materials
hundred charts

Pre-Assess
Give each child a hundred chart. Say: **Skip-count by twos beginning with 2. Shade the square for each number you skip-count.** Say: **Now skip-count by fives beginning with 5. Circle the square for each number you skip-count.** Say: **Now skip-count by tens beginning with 10. Draw an X over the square for each number you skip-count.** Check children's work.

Common Misconception
- Children may incorrectly begin with the number 1 when skip-counting on a hundred chart.

- To correct this, point out that if they are not asked to start on a specific number, then they should begin at zero as they do on a number line. Also point out that since zero is not shown on the hundred chart children should mark the first number they say as they skip-count.

Learn the Math
Review counting by twos, fives, and tens with children before working through the problem on student page **IIN103** with children.

Talk Math
- **Look at the hundred chart. What number pattern do you see in the shaded boxes?** Each number is 2 more than the number before it. The numbers all have a 2, 4, 6, 8, or 0 in them.

- **In the first problem, from what number are you asked to start skip-counting?** 24

Name_____

Learn the Math

Skip-Count on a Hundred Chart

Skill 52

You can use a hundred chart to skip-count.

1	2	3	4	5	6	7	8	9	10
11	12	13	14	15	16	17	18	19	20
21	22	23	24	25	26	27	28	29	30
31	32	33	34	35	36	37	38	39	40
41	42	43	44	45	46	47	48	49	50
51	52	53	54	55	56	57	58	59	60
61	62	63	64	65	66	67	68	69	70
71	72	73	74	75	76	77	78	79	80
81	82	83	84	85	86	87	88	89	90
91	92	93	94	95	96	97	98	99	100

Use the hundred chart to skip-count.
Start at 24. Skip-count by twos.

24, 26, 28, 30, 32, 34, 36

Start at 20. Skip-count by fives.

20, 25, 30, 35, 40, 45, 50

Intensive Intervention • IIN103

Skip-count aloud by twos with children from 24 to 36.

Direct children to the second problem. Explain that this problem is different because they will be skip-counting by fives.

- **What do you notice about the numbers that are circled on the hundred chart?** Each number is 5 more than the one before it. They all end in 0 or 5.

- **From what number are you asked to start skip-counting?** 20

Indicate 20 on the hundred chart. Point and pause at each number on the chart as you skip-count aloud with children.

Do the Math — Skill 52

Use the hundred chart to skip-count.

1. Start at 15. Skip-count by fives.

15, 20, 25, 30, 35, 40, 45

2. Start at 42. Skip-count by twos.

42, 44, 46, 48, 50, 52, 54

3. Start at 30. Skip-count by fives.

30, 35, 40, 45, 50, 55, 60

4. Start at 20. Skip-count by tens.

20, 30, 40, 50, 60, 70, 80

IIN104 · Intensive Intervention

© Houghton Mifflin Harcourt

Alternative Teaching Strategy

Colorful Patterns

Objective To skip-count by twos, fives, and tens on a hundred chart

Materials hundred charts

• Give each child 3 hundred charts.

• Say: **Use a hundred chart to count by twos to 100. As you count, make a mark on each number you say.**

• Say: **Now choose a color and shade each square you marked.**

• Say: **Next, skip-count by fives on another hundred chart. Make a mark on each number you say. Then choose another color to shade the squares you marked.**

• Say: **On the third hundred chart, skip-count by tens. Mark each square of a number you say. Then use a different color to shade the squares you marked.**

• Have children compare their hundred charts side by side and discuss how they are different.

Do the Math

Explain to children that they will use the hundred chart on student page **IIN104** to skip-count by twos, fives, and tens.

Talk Math

• **When you skip-count by tens, what is the last digit of every number?** zero

• **Do you count more numbers when you skip-count by fives or tens to 100?** fives **Why?** It is a lesser number so you have to stop on more numbers.

Check

Say: **Tess has 4 groups of beads with 5 beads in each group. What number will you skip-count by to find how many beads Tess has in all? Explain.** You skip-count by fives because there are five beads in each group. Since there are four groups you skip-count four times: 5, 10, 15, 20. Ask: **How many beads does Tess have in all?** 20